HANDBOOKS
FOSSILS

DK SMITHSONIAN HANDBOOKS

FOSSILS

DAVID J. WARD

Photography by
COLIN KEATES ABIPP

DAVID J. WARD FGS

NEW EDITION

DK LONDON

Managing Editor Angeles Gavira
Managing Art Editor Michael Duffy
US Editor Kayla Dugger
US Executive Editor Lori Cates Hand
Production Editor George Nimmo
Senior Production Controller Meskerem Berhane

DK DELHI

Senior Editor Janashree Singha
Project Editor Hina Jain
Art Editor Shipra Jain
Illustrator Priyal Mote
Managing Editor Soma B. Chowdhury
Senior Managing Art Editor Arunesh Talapatra
Senior Jacket Designer Suhita Dharamjit
Senior DTP Designer Harish Aggarwal
Jackets Editorial Coordinator Priyanka Sharma
Managing Jackets Editor Saloni Singh
Production Manager Pankaj Sharma
Pre-production Manager Sunil Sharma
DTP Designers Syed Md Farhan, Vikram Singh
Editorial Head Glenda Fernandes
Design Head Malavika Talukder

Scientific Editor Alison E. M. Ward

This American Edition, 2021
First American Edition, 1992
by Dorling Kindersley Limited
Published in the United States by DK Publishing
1450 Broadway, Suite 801, New York, NY 10018

Copyright © 1992, 2000, 2021 Dorling Kindersley Limited
Text copyright © 2021 David J. Ward
DK, a Division of Penguin Random House LLC
21 22 23 24 25 10 9 8 7 6 5 4 3 2 1
001–322113–Oct/2021

A catalog record for this book
is available from the Library of Congress.
ISBN 978-0-7440-3000-6

Printed and bound in China

For the curious
www.dk.com

Contents

AUTHOR'S INTRODUCTION

Fossil collecting is a fascinating hobby which has grown considerably in popularity over the last few decades. Its appeal is understandable; it combines the excitement of discovery with the practical skills of collecting and preparing specimens and the academic challenge of identifying fossil finds. There are few other branches of science in which a beginner can make a serious contribution to the knowledge of our planet's remarkable prehistory.

Originally, the word "fossil" (derived from the Latin word *fossilis*, meaning "to be dug up") referred to anything that had been buried. It included not only the petrified remains of plants and animals, but also rocks, minerals, and man-made artifacts, such as coins. It is now used only to refer to the naturally buried and preserved remains of organisms that lived long before historic times.

YEARS OF SPECULATION

Fossils have intrigued people for generations. Greek philosophers regarded them as rather strange, natural phenomena, which formed in the earth, in a similar way to a stalactite or crystal. Martin Luther (1483–1546) believed that fossil finds on mountain tops were evidence of the biblical Flood. In his notebooks, Leonardo da Vinci (1452–1519) suggested that fossils were the petrified remains of once-living organisms. His views, heretical for his era, were withheld until his notebooks were published in the 19th century. The true nature of fossils became slowly

Trachyphyllia (coral)

apparent in the 17th and 18th centuries. This was aided by the publication of books figuring collections of fossils, and by a wider understanding of natural history. One key observation, in the early 19th century, was that the sequence of sedimentary rocks in a region was essentially the same over a large distance and that each layer could be recognized by its own suite of fossils. This allowed rocks to be identified and traced over large distances and was of considerable help in the production of the first geological maps. This led to the modern sciences of paleontology (the scientific study of fossils) and stratigraphy (the study of rock strata). Today, paleontology is concerned only with the remains of animals and plants that lived more than 10,000 years ago.

CLASSIFICATION

Fossils are usually referred to by their two-part scientific name, although a few have popular or informal names as well.

Diplomystus (fish)

***Hemicidaris*
(sea urchin)**

moved to a different genus. When correctly used, a scientific name refers only to a single type of organism and can be understood by scientists all over the world. The basic unit of classification is the species. There are many and varied definitions, but essentially all members of a species look similar and are able to interbreed.

The definition of a species is based on the combination of a detailed written description; illustrations; and, where practical, a specimen or group of specimens preserved for posterity in a museum or similar institution. The key specimen used to identify others of the same species is referred to as the Holotype.

One or more species may be grouped into a genus, linked by features they share. This, along with the family (a group of genera) and the order (a group of families), makes up the pedigree or family tree of an organism. All the stages above "species" are artificial, a man-made classification, and they tend to change depending on current opinion. This can be frustrating for both the beginner and the specialist.

For instance, the oyster *Gryphaea* is often called a "Devil's Toenail," and brachiopods are known as "lamp shells." These names have their uses but lack the precision needed in science; more importantly, they are not internationally accepted and can be confusing. The usual form is to give the full scientific name, usually written in italics, followed by the name of the author, the person who first described the species. The first part is the genus, the second part is the species. If the author's name is in brackets, it means that the species has, at a later date, been

Field trip: January 1988
The author, David Ward *(right)*, with museum curator, Cyril Walker *(left)*, in the southern Sahara, examining an exciting fossil find: a scatter of dinosaur bones. Some of the bones found can be seen on p.249.

AIMS AND LIMITATIONS

This book is intended to assist the collector by illustrating a broad range of fossils, from those most likely to be found, to some of the more spectacular but less common. The fossils were chosen from the Natural History Museum, London, UK, one of the largest and most diverse collections in the world, as well as favorites from the author's personal collection. Microscopic specimens have not been included; although many are fascinating and visually stunning, their study is quite specialized. Most major groups of fossils are included, from worms to dinosaurs, from ammonites to humans, and from all geological ages and continents. The selection and description of each fossil has been assisted by experts who work on the many different types of fossils included. Technical terms have been kept to a minimum, but where this has proved difficult, they have been explained in a comprehensive glossary (see pp.312–316).

Many fossils, particularly the larger reptiles and mammals, are only occasionally found whole. This poses a problem in terms of identification. In such cases, small parts of the skeleton have been illustrated.

Crinoid stem

Amber (fossilized plant resin)

It would be impossible to show a photograph of every type of fossil. However, the range and diversity of specimens contained in this book should enable the collector to find a photograph and description of something sufficiently close to attempt a preliminary identification. In this revised and updated edition, we have attempted to include some of the fossils that are now widely available in rock and mineral stores and that are both legally and sustainably sourced.

Homo habilis (skull of early human)

Calliostoma (gastropod)

HOW THIS BOOK WORKS

THE MAIN BODY OF THE BOOK is divided into four parts: invertebrates, vertebrates, algae, and plants. Within each division, the main groups are introduced. The genus is the starting point of identification. Usually, a typical or relatively common species of the described genus is illustrated. Identifying features are clearly highlighted in annotation.

Occasionally, an unusual specimen, but one that makes an interesting point, has been chosen. Each photograph is accompanied by a reconstruction of a typical species of the organism in life. Some of the reconstruction details, such as the color, are educated guesswork. This annotated example shows a typical page.

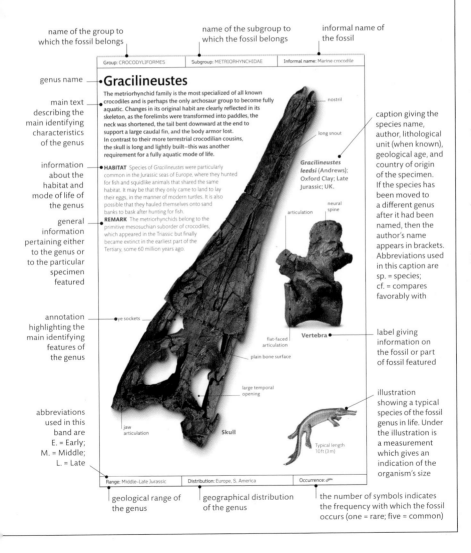

name of the group to which the fossil belongs

name of the subgroup to which the fossil belongs

informal name of the fossil

Group: CROCODYLIFORMES   Subgroup: METRIORHYNCHIDAE   Informal name: Marine crocodile

genus name

Gracilineustes

main text describing the main identifying characteristics of the genus

The metriorhynchid family is the most specialized of all known crocodiles and is perhaps the only archosaur group to become fully aquatic. Changes in its original habit are clearly reflected in its skeleton, as the forelimbs were transformed into paddles, the neck was shortened, the tail bent downward at the end to support a large caudal fin, and the body armor lost. In contrast to their more terrestrial crocodilian cousins, the skull is long and lightly built—this was another requirement for a fully aquatic mode of life.

information about the habitat and mode of life of the genus

HABITAT Species of *Gracilineustes* were particularly common in the Jurassic seas of Europe, where they hunted for fish and squidlike animals that shared the same habitat. It may be that they only came to land to lay their eggs, in the manner of modern turtles. It is also possible that they hauled themselves onto sand banks to bask after hunting for fish.

general information pertaining either to the genus or to the particular specimen featured

REMARK The metriorhynchids belong to the primitive mesosuchian suborder of crocodiles, which appeared in the Triassic but finally became extinct in the earliest part of the Tertiary, some 60 million years ago.

nostril

long snout

Gracilineustes leedsi (Andrews); Oxford Clay; Late Jurassic; UK.

neural spine

articulation

caption giving the species name, author, lithological unit (when known), geological age, and country of origin of the specimen. If the species has been moved to a different genus after it had been named, then the author's name appears in brackets. Abbreviations used in this caption are sp. = species; cf. = compares favorably with

annotation highlighting the main identifying features of the genus

eye sockets

flat-faced articulation

plain bone surface

Vertebra

label giving information on the fossil or part of fossil featured

large temporal opening

abbreviations used in this band are E. = Early; M. = Middle; L. = Late

jaw articulation

Skull

Typical length 10ft (3m)

illustration showing a typical species of the fossil genus in life. Under the illustration is a measurement which gives an indication of the organism's size

Range: Middle–Late Jurassic   Distribution: Europe, S. America   Occurrence:

geological range of the genus

geographical distribution of the genus

the number of symbols indicates the frequency with which the fossil occurs (one = rare; five = common)

WHAT IS A FOSSIL?

FOSSILS ARE THE REMAINS of long-dead plants and animals that have partly escaped the rotting process and have, after many years, become part of the Earth's crust. A fossil may be the preserved remains of the organism itself, the impression of it in the sediment, or marks made by it in life (known as a trace fossil). For fossilization to occur, rapid burial, usually by water-borne sediment, is required. This is often followed by chemical alteration, where minerals may be added or removed. The replacement by and/or addition of minerals usually aids preservation.

Fossil dung
Trace fossils are true fossils. This dropping (below left) is probably from an extinct shark. It is usually difficult to relate coprolites to the animal that produced them.

Reptile footprint
While difficult to identify to a particular species, fossil footprints (above) can provide valuable information about the organism's behavior, such as speed, weight, and mode of life.

Horse tooth
This cheek tooth (below) looks like that of a modern horse, but it is actually a fossil. Over the centuries, its organic tissue has been replaced by strengthening mineral salts, ensuring its preservation.

Mummified frog
Mummification—the natural drying of an organism—has meant that the frog (right) has progressed part of the way toward becoming a fossil. To ensure proper preservation, it must become entombed in a medium that would guard against further decomposition.

Banded flint
Flints (right) are sometimes mistaken for fossils. During formation, flint can be deposited in bands. With a little weathering and staining, these flints can resemble fossil corals, mollusks, worms, and trilobites.

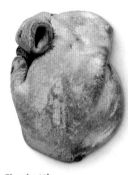

Foot-shaped
Cretaceous flints come in many forms; this one resembles a human foot (below). Some flints are crustacean burrow infills; if so, they are regarded as trace fossils.

Birds' nest
The birds' nest (above) is not a fossil; it is a modern nest that has been petrified in a spring.

Clay bottle
Objects as mundane as a collapsed clay bottle (above) are sometimes mistaken for fossils.

HOW FOSSILIZATION OCCURS

Fossilization is, at best, a risky process which relies on a chain of favorable circumstances. The vast majority of the plants and animals that have ever lived have completely disappeared without a trace, leaving no fossil record. With rare exceptions, it is only the skeletal or hard parts of an organism that become fossilized. This often occurs when the organism decaying in the sediment alters the local conditions and promotes the incorporation of mineral salts within its structure, a process known as mineralization. This chemical change often enables the fossil to become more resistant than the surrounding sediment.

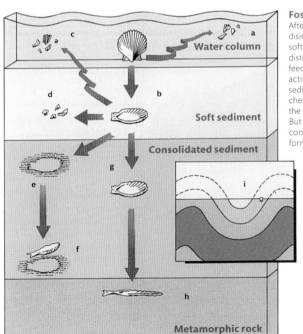

Fossilization
After death, an organism may slowly disintegrate (a) or become buried in soft sediment (b). However, it may be disturbed or digested by sediment-feeding organisms, or current or wave activity may reexpose it (c). As the sediment compacts and the complex chemical reactions of diagenesis occur, the potential fossil may be dissolved (d). But if the sediment is sufficiently consolidated, a mold may be formed (e). Percolating mineral solutions may infill the mold, creating a cast (f). Some enter the sediment relatively unaltered by mineralization (g). If buried and subjected over time to increased temperature and pressure, sedimentary rocks become softened and distorted (metamorphosed), and ultimately destroyed (h). As rocks are folded, uplifted, and eroded, buried fossils may be exposed on the surface (i).

MODES OF PRESERVATION

TO BECOME preserved as a fossil, some of the normal processes of decay must be permanently arrested. This usually involves isolating the organisms that cause decay from the air or water and then filling any voids in the hard tissue with additional minerals. The vast majority of fossils are, therefore, found in freshwater or marine sediments, where oxygen-deprived silt or clay has buried the organism soon after death. If the sediment conditions remain favorable (see pp.10–11), the organism may be preserved as a fossil. In the case of mummification, the arrest of decay is only temporary; a mummified organism will begin to decay as soon as it is exposed to air once again.

Under exceptional circumstances, soft-tissue details may be preserved. Insects in amber and mammoths in ice or tar are well-known examples. In both these cases, the living organism has been caught in the sticky substance (tar or resin) which has then been fossilized, ensuring preservation. If limestone, phosphate, or pyrite is deposited in the sediment surrounding a decaying plant, it forms a "tomb" that may preserve very fine details of the organism. Silicified or petrified wood can produce spectacular color effects, although the cell preservation itself is often poor.

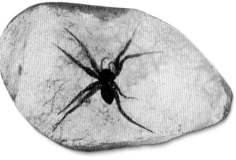

Trapped in amber
Amber, the fossilized resin of a plant or tree, sometimes preserves not only the external, but also the internal, structure of an organism. Insects, spiders (above), frogs, and lizards may be preserved in this way.

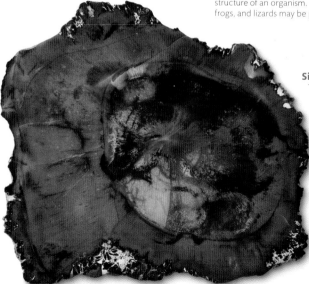

Silicification
This is a form of "petrification." Silicified wood (left) can often be found in both terrestrial and freshwater deposits, usually sands and silts. Weathering volcanic ash usually supplies silica, which is gradually incorporated into partially decayed wood. Generally, the cell structure of silicified wood is quite poorly preserved; however, the presence of iron and other minerals can produce some really spectacular coloring.

Phosphatization

Bones and teeth (above), which normally dissolve on the sea bed or leach from the sediment, are more likely to be preserved if large quantities of phosphates are present. Phosphatic deposits are a source of well-preserved fossils and are often mined commercially.

Mummification

The dry, sterile atmosphere of a cave has preserved this moa foot (above) with some soft tissue intact; usually bones are only preserved in a fragile state. Mummification is only a pause in disintegration and is not true fossilization.

Freezing

Siberian mammoths are often found preserved in permafrost. Once thawed, they will decay unless action is taken, but the hair is relatively durable.

Tar and sand

A mixture of tar and sand has embalmed this beetle (left). This could be stable for thousands, but not millions, of years.

Limestone tomb

The completeness of this fragile crinoid, entombed in limestone (left), suggests that the calcareous nodule encasing it formed soon after death.

Pyritization

The shell and chambers of this ammonite (right) were replaced by iron pyrite. This is often unstable in moist air, so it needs to be stored in very dry conditions.

GEOLOGICAL TIME CHART

THE PLANET EARTH was formed 4,600 million years ago, with life present for at least 3,850 million years. Although multicellular life appeared over 1,000 million years ago, remains from that period are scarce. The first organisms with hard parts, allowing fossils to become relatively common, appeared about 550 million years ago. Geological time is divided into periods, usually named after the area where the rocks from that period were first exposed:

for example, Jurassic is named after the Jura Mountains between France and Switzerland; Devonian after the rocks in Devon, UK. These periods provide a framework which, while estimates of their actual age may alter quite dramatically, allow rocks and fossil remains to be correlated worldwide.

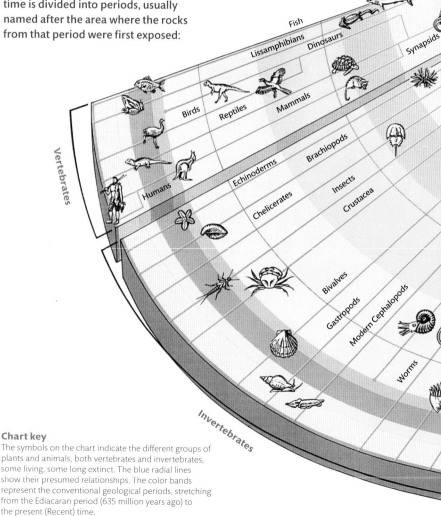

Chart key

The symbols on the chart indicate the different groups of plants and animals, both vertebrates and invertebrates, some living, some long extinct. The blue radial lines show their presumed relationships. The color bands represent the conventional geological periods, stretching from the Ediacaran period (635 million years ago) to the present (Recent) time.

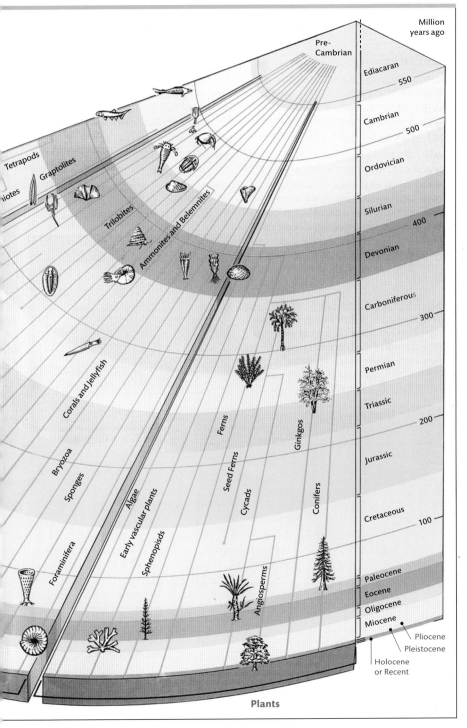

Million years ago

Pre-Cambrian

Ediacaran — 550

Cambrian — 500

Ordovician

Silurian — 400

Devonian

Carboniferous — 300

Permian

Triassic — 200

Jurassic

Cretaceous — 100

Paleocene

Eocene

Oligocene

Miocene

Pliocene

Pleistocene

Holocene or Recent

Tetrapods

Graphtolites

Trilobites

Ammonites and Belemnites

Corals and Jellyfish

Bryozoa

Sponges

Foraminifera

Algae

Early vascular plants

Sphenopsids

Ferns

Seed Ferns

Cycads

Ginkgos

Conifers

Angiosperms

Plants

WHERE TO LOOK

YOU CAN FIND FOSSILS in most places where sedimentary rocks, such as clays, shales, and limestones, are exposed. Exposures of hard rock recover very slowly from collecting and can be disappointing. Artificial exposures, like road cuts or quarries, can be much more productive. Exposures of softer rocks can be good sites, provided they have not been too badly altered by metamorphism. Inland sections in sands and clays tend to degrade rapidly, become overgrown, and lost. Here temporary exposures are valuable, along with continually eroding river or coastal sections.

Establishing the age of the rocks, with the aid of a geological map, will give you an idea of what sort of fossils to expect. Most libraries have geological guides, but remember that they may be out of date: many an hour has been spent looking for quarries that have been infilled and had houses built on them.

A visit to a local museum is often useful, but up-to-date information from other collectors is better. Consider joining a natural history society or a rock and mineral club. They can often gain access to private localities. Most geological societies have a code of conduct relating to collecting, which it is advisable to follow. Wherever you decide to collect fossils, be aware of local regulations and obtain permission from the owner.

Coastal sites
Wave-washed cliffs and foreshore exposures (above) are good places to search for fossils. Be aware of the state of the tide when you are rounding headlands. Wear a hard hat to protect you from small stones dislodged by startled seabirds, but remember it will not protect you from larger falling rocks.

Quarries
Supervised parties are usually allowed to collect fossils in quarries (left), but individuals may be discouraged. Check beforehand whether there are places, usually around active faces, that are out-of-bounds. Hard hats are a normal requirement. The staff often know where the best specimens may be found.

Trilobite
Concoryphe (left), an eyeless deep-water genus, was found in a hard Cambrian mudstone.

Arid terrains
Although erosion is slow in deserts (below) and badlands, the large areas of exposure that exist can more than compensate.

National parks
Hammering rocks and fossil collecting are illegal in national parks (above), so please be sensitive. In this way, fossils will still be there for future generations.

Rugged terrain
Fossils are often concentrated along particular bedding planes; the combination of a potentially interesting area and well-stratified rocks make an ideal hunting ground (below).

COLLECTING FOSSILS

COLLECTING FOSSILS is a very relaxing and intellectually stimulating hobby, but it is one that can often be frustrated by not having suitable field equipment. It is clearly impossible to cater for all eventualities, but the tools shown below form a basic selection that covers most situations. However, it is always better to leave a fossil in the field than to try to dig it out in a hurry with the wrong tools; this could damage a valuable scientific find.

Occasionally, equipment for plastering is required. You use this to protect fragile fossils before you remove them from the rock. Clean the fossil and expose it as much as possible. Cover it with layers of a separator (wet paper or plastic wrap), followed by layers of plaster bandage. Once the plaster has been set, you can lift the fossil out of the rock face, then repeat the process on the underside.

Your safety should be a primary concern. Hard hats, goggles, and gloves are essential. Finally, before you set out on an expedition, ask other collectors, who are familiar with the area you intend to visit, exactly what tools you will need.

compass

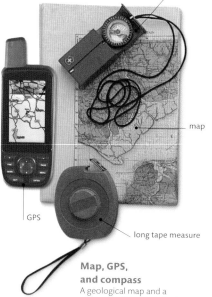
map
GPS
long tape measure

SAFETY EQUIPMENT

Rocks can be sharp and dangerous. A hard hat is essential if collecting near high rock faces. Goggles will shield your eyes from dust and stones, and gloves will protect your hands.

Map, GPS, and compass
A geological map and a compass will help you find fossil localities. Take a long tape measure to record the level of the bed in which you find fossils.

Notebook and smartphone
Record field notes, such as locality, type of rock, and fossils seen, in a sturdy notebook in waterproof ink. A photograph can also prove invaluable.

Field tools

For extracting fossils from hard rocks, a sturdy mallet and several guarded chisels are essential equipment. A pick-ended geological hammer is useful on most types of rock. Soft sediments are best tackled with a trowel or a spade.

mallet

geological hammer

trowel

hand lens

guarded chisels

Sieve

Use a sieve to separate fossils from sands and gravels. Usually one or two different meshes are needed to avoid losing small fossils.

Plastering

Protect fragile fossils in a plaster jacket before removing them from the rock. This prevents them from shattering.

plaster bandages

small brush

large brush

plastic sieve

Spade

When you are searching for fossils in soft sediments—such as sands, silts, and clays— a narrow-bladed spade, rather than a conventional geological hammer, is the most useful tool for clearing an area around the fossil.

PREPARING YOUR COLLECTION

UNLIKE MOST OUTDOOR activities, fossil collecting does not end when you return from a field trip. A freshly collected specimen usually needs cleaning to remove any adherent matrix. This can be a very simple procedure, such as dusting sand off a specimen, or it may involve using dilute acids to remove fossils encased in hard rock. For example, you will need acid to remove fossilized teeth from an unconsolidated matrix of shell debris. This is a tricky process that should be first attempted with the guidance of an expert.

Fragile fossils can be strengthened with a dilute solution of consolidant. Use a glue that can be removed so that a joint can be repositioned later. Plastic boxes are ideal for storing small fossils. Open cardboard trays suit larger specimens. Use glass vials, microscope slides, or even gelatin capsules to house very small specimens.

For best practice, each fossil should have a catalog number but also be housed with its own label as a backup. The information can be stored in a card index file or a computer catalog. The label and catalog should include the following data: the name, the lithological unit, and the rock's geological age. This should be followed by the locality, its map reference, the country or state, and the date it was collected. A fossil without these details is of little scientific value.

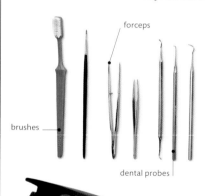

brushes

forceps

dental probes

Cleaning fossils
Use a variety of large and small brushes and dental probes to clean the specimen and to remove small fragments of rock. Chip off more resistant matrix with a lightweight hammer and chisel or with an engraving tool. Handle small or fragile specimens very carefully, preferably with fine forceps. It is a good idea to protect them with a consolidant so that they will not break up when they are handled or transported.

dilute acid

engraving tool

air scribe

sieve

sharks' teeth

Sieving
Use a stack of stainless-steel or brass sieves to separate samples containing small fossils, such as sharks' teeth, into different sizes. Brass sieves are soft and damaged by sea water, and so are not suitable for use in the field.

Organic acids
Use dilute organic acids— for example, vinegar—to remove the surplus matrix from a calcareous specimen. Practice the procedure on an unimportant fossil first to avoid damaging a valuable specimen.

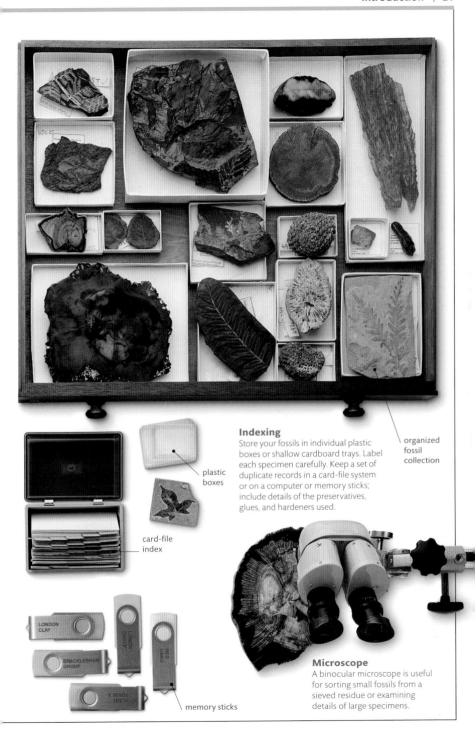

Indexing
Store your fossils in individual plastic boxes or shallow cardboard trays. Label each specimen carefully. Keep a set of duplicate records in a card-file system or on a computer or memory sticks; include details of the preservatives, glues, and hardeners used.

organized fossil collection

plastic boxes

card-file index

LONDON CLAY

LOWER SHALE?

BRACKLESHAM GROUP

RED CRAG

PLANT FOSSILS

memory sticks

Microscope
A binocular microscope is useful for sorting small fossils from a sieved residue or examining details of large specimens.

FROM HOBBY TO SCIENCE

BEGINNERS NORMALLY START by collecting any fossil that comes their way. This is not a bad idea, as there is a lot to learn about the hobby. After a while, however, most collectors specialize in a particular group of fossils, such as trilobites, ammonites, or plant fossils, or perhaps an age or locality.

Initially, most collectors tend to regard fossils merely as objects that are attractive in their own right, a pleasure to admire and own. With time and experience, the specimens themselves become less important and collectors soon realize that their true value lies in the scientific information they represent. Donating specimens to a local or national museum collection is one way a fossil collector can repay any help that he or she has received from museum staff in the past; most museums rely very heavily on the generosity of amateur collectors.

It is no exaggeration to say that paleontology is one of the few sciences where an amateur is able make a significant contribution to the state of knowledge—in fact, this is probably one of its major attractions as a hobby.

National museums
These often have an identification service and good regional and national exhibits of fossils (left), allowing collectors to make their own determinations. Only the best of a vast collection is on display, so try not to be discouraged if your fossils don't quite match in quality. The museum shop will usually have a good selection of books on fossil identification.

Local museums
Most museum curators are happy to give advice, discuss the local geology, and help identify problem specimens (right). Their knowledge and enthusiasm has inspired many young collectors to pursue geology as a career. Local museums can also put you in touch with the nearest geological society.

Easy find
Shark's teeth are a favorite among fossil collectors. In rocks of the right age, they are quite easy to find.

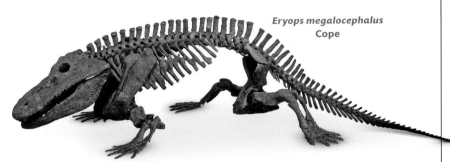

Eryops megalocephalus
Cope

Most fossil collectors dream of finding something new. This need not be a large and spectacular specimen, like a dinosaur; it is more often a small and apparently insignificant fossil. The skill, for the amateur paleontologist, lies in recognizing something unusual. If it is a new species, a description can be published in a scientific journal, accompanied by a photograph and details of the locality—a procedure that can take several years. The specimen should ideally be housed in a museum collection, which ensures that all those interested can study it. A private collection is not usually acceptable if it is not open to the public. Parting with a prize specimen can be difficult, but it is necessary if the full potential of the discovery is to be realized.

Museum exhibit
This giant amphibian (above) is from the Permian of Texas. Magnificent specimens like this, reconstructed by paleontologists, are a feature of natural history museums worldwide.

BUYING FOSSILS

You can purchase small fossils, such as mollusks and trilobites, quite cheaply in rock and mineral shops, at fossil shows, and on the internet. Collecting, owning, and exporting fossils is illegal in many countries, although possessing them outside of that country may not be an offense. It is important to respect such laws by ensuring that all potential purchases have been legally exported.

Reference books
Libraries may house old illustrated monographs. Paleontology is one of the few branches of science in which something written over a century ago can still be relevant.

FOSSIL IDENTIFICATION KEY

THE FOSSIL RECORD is so vast that to give adequate coverage in a book this size is impossible. Nevertheless, typical examples of most of the more common groups of fossils have been illustrated here. In this way, even if a particular genus is not illustrated, a similar organism will be shown. Over the next eight pages, visual examples of each major group of fossils are given, along with a concise description and the pages of the book where the group is covered in detail. When attempting an identification, it is vital to relate the range of any fossil to the rock in which your particular specimen is found; for instance, you do not find trilobites in the Cretaceous, nor mammals in the Devonian. A look at the geological time chart on pages 14–15 will help you with this. The key is divided into three sections: invertebrates, vertebrates, and plants.

INVERTEBRATES

FORAMINIFERANS

Small (some microscopic), multichambered, calcareous objects; either round, discoidal, or ovoid. Usually present in large numbers, they can be rock-forming.
Range Cambrian–Recent

Page: 32

Test of *Nummulites*

WORMS

Burrows, tracks, or calcareous tubes; usually spiral, cemented to objects. Can be confused with teredinid tubes (p.111), scaphopods (p.114), or corals (p.50).
Range Cambrian–Recent

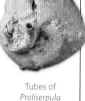

Tubes of *Proliserpula*

Pages: 40–41

Tubes of *Rotularia*

SPONGES AND BRYOZOANS

Composed of calcite, silica, or as an outline on a bedding plane. Shaped more like a plant than an animal; irregular, often branching, treelike; size variable. Sponges are generally thick-walled, with a featureless or coarsely ornamented surface. Bryozoa are usually thin-walled, with a delicate pore network; they can encrust other fossils or pebbles. They can be confused with corals (p.50) or calcareous algae (pp.286–288).
Range (sponges) Cambrian–Recent
Range (bryozoa) Ordovician–Recent

Skeleton of *Rhizopoterion*

Pages: 33–39

Skeleton of *Raphidonema*

Skeleton of *Constellaria*

TRACE FOSSILS AND PROBLEMATICA

These encompass tracks, trails, borings, and burrows. Generally they are radiating, feeding, or tubular burrow structures made by worms, arthropods, echinoids, mollusks, or they are walking tracks of arthropods. Terrestrial, shallow lake sediments may show regularly spaced depressions made by "footprints." The systematic position of problematicans is uncertain because they are either poorly preserved or they resemble no modern fossil group.
Range Precambrian–Recent

Footprint of
Tetrapod

Pages: 42–44

Trace of
Chondrites

Imprint of
Spriggina

GRAPTOLITES

These are impressions on rocks resembling watch springs, fret saw blades, or meshwork. "Teeth" may be on one or both sides of stipe, which may be single or multiple. Graptolites may be confused with some plant remains (pp.286–311), or bryozoa (pp.36–39), or an oblique section of crinoid stem (p.168).
Range Ordovician–Carboniferous

Skeleton of
Retiolites

Pages: 45–49

Skeleton of
Didymograptus

Skeleton of
Phyllograptus

CORALS

Shapes are very variable: hornlike, tubelike, or treelike. The skeleton is calcareous. There are one or many calices; each individual calice is divided by a series of radial plates, giving a starlike appearance. Calices may be closely abutted or laterally fused in meandering chains (like brain coral, p.54). The size of the colony can vary from less than a centimeter to several feet. They can be confused with encrusting bryozoa (p.36), calcareous algae (p.287), serpulid worm tubes (p.40), or sponges (p.33).
Range Ordovician–Recent

Colony of
Colpophyllia

Pages: 50–55

Colony of
Favosites

Calice of
Trachyphyllia

ARTHROPODS

Arthropods have segmented bodies with a differentiated anterior end. The body is covered by a cuticle, or shell, that acts as an external skeleton (exoskeleton). The exoskeleton is sometimes strengthened by calcium carbonate or calcium phosphate. Arthropods have a variable number of appendages (legs), which are specialized into locomotory, sensory, or feeding elements. Molting is a common feature of arthropods—otherwise, their growth would be constrained by the external cuticle. Arthropods include the Trilobites, the Crustacea (crabs, lobsters, barnacles, and shrimps), the Chelicerates (king crabs, scorpions, and spiders), and the Uniramia (millipedes and the Insects).

TRILOBITES

These possess a flattened, mineralized, calcium carbonate shell and are preserved in three dimensions or as an impression. The eyes are often fragmented. The thorax may segment.
Range Cambrian–Permian

Exoskeleton of *Phacops*

Internal mold of *Olenellus*

Pages: 56–65

CHELICERATES

The body is divided into a fused head, thorax, and abdomen. They can be confused with some trilobites and insects. Rarely found as fossils.
Range Cambrian–Recent

Imprint of *Mesolimulus*

Body of *Dolomedes*

Pages: 73–75

CRUSTACEANS

Usually the ornamented phosphatic carapace is preserved; sometimes limb fragments, calcareous plates, or bivalved shells are found.
Range Cambrian–Recent

Imprint of *Hymenocaris*

Pages: 66–72

Exoskeleton of *Archaeogeryon*

INSECTS

These are divided into head, thorax, and abdomen. The adults have six legs and sometimes wings.
Range Devonian–Recent

Exoskeleton of *Hydrophilus*

Pages: 76–78

Imprint of *Petalura*

BRACHIOPODS

The shells have two symmetrical, calcareous, or chitinous, dissimilar valves. One valve usually bears a hole (pedicle foramen), through which it is attached to the substrate. They are common in older rocks, often in large "nests" of a single species.
Range Cambrian–Recent

Pages: 79–93

Shell of *Lingula*   Shell of *Goniorhynchia*   Shell of *Terebratula*

MOLLUSKS

Mollusks are a very diverse group that includes the chitons, scaphopods, bivalves, gastropods, and cephalopods. Nearly all mollusks have calcium-carbonate shells; only a few genera of gastropods, cephalopods, and nudibranchs lack this. The soft body is rarely preserved, so mollusks are classified by their shell structure. Most are marine; only a few families of bivalve and gastropod have entered fresh water. Some gastropods have become air breathing and colonized the land. They can be filter feeders (most bivalves), herbivores (some gastropods), or carnivores (gastropods, cephalopods). One major group, the ammonoids, became extinct at the end of the Cretaceous.

BIVALVES

Two valves, similar in shape but each inequilateral; exceptions are oysters and rudists. May be loose in the sediment or cemented to a firm substrate. May be confused with brachiopods (pp.79–93) and corals (p.50).
Range Cambrian–Recent

Pages: 94–113

Cast of *Whiteavesia*

Shell of *Crassatella*

CHITONS AND SCAPHOPODS

Chitons: ridged, tilelike plates. Scaphopods: curved; tapering.
Range (Chitons) Cambrian–Recent *Range* (Scaphopods) Ordovician–Recent

Valves of *Helminthochiton*

Page: 114

Shell of *Dentalium*

GASTROPODS

Single, asymmetrical, spiral, unchambered calcium-carbonate shell.
Range Cambrian–Recent

Pages: 115–133

Shell of *Turritella*

Shell of *Ecphora*

NAUTILOIDS

Straight, curving, or loosely or lightly coiled calcium-carbonate shells, divided into a chambered phragmocone and a living chamber. The chambers are connected by a tube (siphuncle).
Range Cambrian–Recent

Pages: 134–140

Cast of *Orthoceras*

Section of unknown *nautilus*

AMMONOIDS AND AMMONITES

Similar to nautiloids but with a ventral siphuncle. Chambers give rise to a complex suture. Ribs and tubercles are common.
Range Devonian–Cretaceous

Pages: 141–160

Cast of *Mantelliceras*

BELEMNITES AND SQUIDS

These have a solid, tapering, crystalline calcium-carbonate guard, partially enclosing a chambered phragmocone.
Range Jurassic–Recent

Pages: 161–165

Guard of *Belemnitella*

ECHINODERMS

The echinoderms include the echinoids (sea urchins), holothuroids (sea cucumbers), asteroids (starfish), crinoids (sea lilies), as well as the extinct blastoids, cystoids, and carpoids. All have five-rayed symmetry except for some echinoids and carpoids. They possess a complex water vascular system that operates a network of multifunctional tube feet.

All are restricted to fully marine environments and thus act as marine indicators. Most possess solid calcite skeletons that fall apart on death; crinoidal debris can be so abundant as to be rock forming (see p.168). Some irregular echinoids (see *Micraster* p.185) have evolved in a manner that makes it possible for them to be used as zone fossils.

CRINOIDS

Comprise a columnar stem, head (or calyx), and arms (pinnules); some species are stemless. On death, they usually disintegrate to isolated calyx ossicles and round, pentagonal, or star-shaped, crystalline calcite stem segments.
Range Cambrian–Recent

Pages: 166–174

Calyx of
Marsupites

Calyx of
Cyathocrinites

ECHINOIDS

The shell or test is composed of thin, loosely cemented, calcite plates. The test may be subspherical and have five-rayed symmetry, or heart-shaped, or dorso-ventrally flattened with a bilateral symmetry (characterized by irregular echinoids). Plates bear spines that may be short and bristlelike or large and club-shaped.
Range Ordovician–Recent

Test of
Hemicidaris

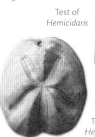

Pages: 175–185

Test of
Hemiaster

ASTEROIDS

These are five-sided, with margins edged with solid calcitic blocks (ossicles). Occasionally preserved as whole, articulated specimens; more usually found as isolated ossicles.
Range Ordovician–Recent

Pages: 186–188

Cast of
Tropidaster

OPHIUROIDS

These small, individual ossicles are common in sieved residues; articulated specimens are rare. The central disk of calcite plates has five, snakelike arms. They are common in deep-water sediments.
Range Ordovician–Recent

Page: 189

Skeleton of
Palaeocoma

BLASTOIDS

Usually intact, compact calyx.
Range Ordovician–Permian

Page: 190

Calyx of
Pentremites

CYSTOIDS

Short stem, calyx of many small, irregular plates.
Range Cambrian–Permian

Pages: 191–192

Calyx and
stem of
Lepadocrinites

CARPOIDS

Stem short, asymmetrical calyx of small polygonal plates, no arms.
Range Cambrian–Devonian

Page: 193

Calyx of
Cothurnocystis

VERTEBRATES

A vertebrate is an animal possessing a flexible segmented spinal column composed of cartilage or bone. The principal modern vertebrate groups are fish, amphibians, reptiles, mammals, and birds. The oldest known fossil vertebrate is from the early Cambrian period.

FISH

Fish are a very diverse group of marine and freshwater vertebrates. All fish possess segmented body musculature, a brain, spinal cord, notochord, terminal mouth, hollow gut, and pharyngeal perforations (usually known as gills). They all have paired pelvic and pectoral fins, unpaired dorsal and anal fins, and a postanal tail.

AGNATHANS

Jawless fish covered with fine scales. The head may be encased in an armored shell of fused scales.
Range Late Cambrian–Late Devonian

Head shield of
Pteraspis

Pages: 194–195

GNATHOSTOMES

Gnathostomes are vertebrates that possess a cranium with a hinged lower jaw.

PLACODERMS

Heavily armored fish with large pectoral fins and sometimes dorsal fin spines. Some have paired bony plates in their jaws.
Range Early Devonian–Early Carboniferous

Pages: 196–197

Skeleton of
Bothriolepis

ACANTHODIANS

Found as isolated scales, teeth, and ornamented fin spines. Fin spines can be confused with those of a shark. Teeth lack enamel.
Range Late Silurian–Early Permian

Page: 210

Skeleton of
Diplacanthus

CHONDRICHTHYANS

Skeletons of sharks, rays, and chimaeroids are rarely preserved articulated. Usually teeth, tooth plates, placoid scales, fin spines, and calcified vertebrae are found. Sharks' teeth have enameloid crowns.
Range Late Silurian–Recent

Tooth of
Striatolamia

Pages: 198–209

Tooth of
Ptychodus

OSTEICHTHYANS

Also known as bony fish, these are found as teeth, bones, scales, fin spines, and otoliths.
Range Late Silurian–Recent

Scales of
Lepidotus

Pages: 211–220

Otolith of
Centroberyx

TETRAPODS

See next page

TETRAPODS

Tetrapods are a group of four-limbed vertebrates that were derived from lobe-finned fish. It includes extant and extinct amphibians, reptiles, dinosaurs, and birds, and the synapsids (including mammals).

Pages: 221–224

LISSAMPHIBIANS

The bones are solid or hollow; the skull is composed of fused bony plates. The teeth are simple with no roots. Vertebrae may be unitary or multipartite and have neural canals. Limb bones may bear a variable number of digits. *Range* Late Devonian–Recent.

Page: 224

Skeleton of *Rana*

AMNIOTES

Amniotes are tetrapod vertebrates that include synapsids, mammals, reptiles, dinosaurs, and birds. They lack the larval stage of lissamphibians and have an amniotic membrane protecting the embryo.

REPTILES

Diverse grouping of tetrapod. Skulls with or without secondary openings. Limb bones are usually hollow (except turtles). Teeth are enameled with a single long root, set in sockets or fixed to jaws. *Range* Late Carboniferous–Recent

Carapace of *Trionyx*

Pages: 225–244

DINOSAURS

Bones are solid, or thick-walled and hollow. Skull is lightly built with fixed quadrate bones; enameled teeth lie in deep sockets. Hip socket is perforated; head of thigh bone is angled inward. *Range* Late Triassic–Cretaceous

Tooth of *Jobaria*

Pages: 245–255

BIRDS

Skull is light, teeth are usually absent. Limb bones are hollow and thin-walled, sternum is keeled, ankle bones are fused. *Range* Late Jurassic–Recent

Limb bone of *Aepyornis*

Pages: 256–259

SYNAPSIDS

Bones are solid, vertebrae biconcave, neural canal is present. Skull is often massive. Teeth differ. *Range* Late Carboniferous–Jurassic

Skull of *Cynognathus*

Pages: 260–262

MAMMALS

Limb bones are usually hollow and thin-walled; epiphyseal plates present. Teeth are enamel-covered and multifaceted, some have bony roots. *Range* Late Triassic–Recent

Tooth of *Tetralophodon*

Skull roof of *Bison*

Pages: 263–285

ALGAE

Preserved as microscopic single-celled plants, banded stromatolitic masses, calcified structures, or thin impressions.
Range Precambrian–Recent

Pages: 286–288

Colony of
Mastopora

PLANTS

EARLY LAND PLANTS AND HEPATOPHYTES

Present as small, upright, branching, leafless, aerial shoots, or low thallus with ovoid spores; no true roots.
Range Late Silurian–Recent

Pages: 289–290

Thallus of
Hexagonocaulon

SPHENOPSIDS

Vertical stems emerge from underground rhizomes. The main stem is jointed, and pith-filled with leafed stems.
Range Early Devonian–Recent

Pages: 290–291

Frond of
Asterophyllites

FERNS, LYCOPODS, AND PTERIDOSPERMS

The fronds of ferns and pteridosperms are divided into multiple leaflets. In lycopods, the leaves are spirally arranged. Spores are borne on the underside of leaves in ferns; in cones in the leaf axil in lycopods. Pteridosperm fronds bear a single seed.
Range Late Silurian–Recent

Cone of
Equisitites

Pages: 292–299

Leaflets of
Dicroidium

Stem of
Osmunda

BENNETTITES AND PROGYMNOSPERMS

Foliage grows from large trunk. Cones are in center of crown or leaf bases; flowers are present.
Range Triassic–Recent

Page: 300

Flower of
Williamsonia

CORDAITALES AND CONIFERS

Woody trunks, often rich in resin; leaves are needlelike, spirally arranged. Male and female reproductive structures present on a single tree; seeds in cones.
Range Late Carboniferous–Recent

Cone of
Picea

Pages: 301–306

GINKGOS AND ANGIOSPERMS

Preserved as wood; leaves; and, very rarely, flowers, where both male and female organs may be present. Seeds are large; leaves are deciduous, (some) broad, with a radiating network of veins, or relatively narrow with parallel veins.
Range Cretaceous–Recent

Pages: 307–311

Leaf of
Araliopsoides

Trunk of
Palmoxylon

INVERTEBRATES

FORAMINIFERANS

FORAMINIFERANS ARE SMALL, single-celled organisms belonging to the kingdom Protista. Some species secrete a shell, called a test, made of calcium carbonate or (less commonly) chitin. Others construct a test by sticking together particles of sand and debris. Most species live in the sea: planktonic foraminiferans float on the surface waters of the oceans, their dead tests drifting down to the sea bed; benthonic foraminiferans live in or on the sea bottom. Both types occur in large numbers, and rocks may form from their remains. The few large species live in warm, shallow waters.

Group: ROTALIIDA	Subgroup: NUMMULITIDAE	Informal name: Nummulite

Nummulites

Circular in outline and biconvex, this large foraminiferan is made up internally of small chambers arranged in a spiral.

HABITAT *Nummulites* was abundant in the shallow, warm seas of the ancient Tethys Ocean.
REMARK Nummulitic limestone was used in the building of the Egyptian pyramids.

Typical diameter
⅝ in (1.5 cm)

tiny internal chambers

circular, biconvex form

Nummulites ghisensis (Ehrenberg); Nummulitic Limestone: Eocene; Egypt.

Range: Paleocene–Oligocene	Distribution: Europe, M. East, Asia	Occurrence: ◉◉◉◉◉

Group: MILIOLIDA	Subgroup: ALVEOLINIDAE	Informal name: Alveolinid

Alveolina

Oval in shape and often large in size, this creature is made up internally of numerous small chambers, which in life may have housed symbiotic algae.

Block of *Alveolina* **limestone**

Alveolina elliptica (Sowerby); Kithar Series; Eocene; India.

HABITAT *Alveolina* lived on the sea bed in warm, shallow waters.
REMARK Where locally abundant, *Alveolina* sometimes formed limestones.

Typical diameter
⁵⁄₃₂ in (4 mm)

oval shape

Range: Eocene	Distribution: Europe, M. East, Africa, Asia	Occurrence: ◉◉◉◉◉

SPONGES

SPONGES ARE SIMPLE, sedentary, aquatic animals. Water passes in through their many surface pores to the central cavity of the saclike body and out through larger holes. The skeleton, where present, is made up of needle-like spicules and is either calcareous, siliceous, or horny. Sponges occur as fossils from Cambrian times onward and were abundant in the Cretaceous. Stromatoporoids are also believed to be sponges, but archaeocyathids are now thought to be an independent phylum.

Group: RETICULOSA	Subgroup: DICTYOSPONGIIDAE	Informal name: Glass sponge

Hydnoceras

The thin-walled, vase-shaped *Hydnoceras* is a fine example of a glass sponge, with an open structure of spicules forming a rectangular meshwork. Bulbous swellings appear along the longitudinal ridges.

HABITAT Living glass sponges are found only in deep water, but they were once common at all depths.
REMARK This fossil is an internal mold in sandstone, preserving the filling of the skeleton but not the skeleton itself. The meshwork is clearly visible, but not the spicules themselves.

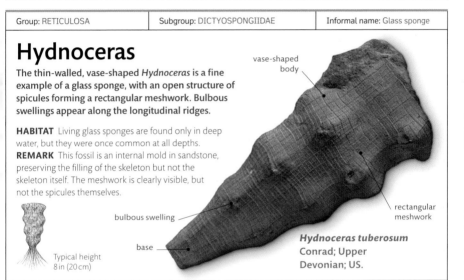

vase-shaped body

rectangular meshwork

bulbous swelling

base

Typical height 8 in (20 cm)

Hydnoceras tuberosum Conrad; Upper Devonian; US.

Range: Late Devonian–Carboniferous	Distribution: Eastern US, Europe	Occurrence: ◉◉

Group: LITHISTIDA	Subgroup: KALIAPSIDAE	Informal name: Calcisponge

Laosciadia

This flat, mushroom-shaped sponge, common in the Late Cretaceous, is a lithistid, a kind of demosponge characterized by a compact, complex structure with many chambers. The skeleton consists of a rigid interlocking framework of four-rayed spicules.

HABITAT These siliceous sponges lived in depths of water varying from 330 to 1,300 ft (100–400 m).

Typical height 3¼ in (8 cm)

Laosciadia plana (Phillips); Upper Chalk; Late Cretaceous; UK.

upper surface

rigid framework of spicules

holdfast

Range: Cretaceous	Distribution: Europe	Occurrence: ◉◉◉◉◉

Group: LYCHNISCOSIDA	Subgroup: VENTRICULITIDAE	Informal name: Glass sponge

Rhizopoterion

This generally funnel-shaped genus is one of the glass sponges, or hexactinellids, which are very different in their organization from other sponges found as fossils. Their siliceous skeletons have an open structure of spicules, with four to six rays mutually at right angles, forming a rectangular meshwork. Within the glass sponges, ventriculitids form an important group, of which *Rhizopoterion* is a member. They vary in form, but all are characterized by the form of their spicules and by having walls pierced by an irregular array of slotlike inhalant and exhalant canals. Their bases have a holdfast of radiating "roots," and they range in shape from tall, narrow vases to flat, open mushrooms.

HABITAT Glass sponges such as *Rhizopoterion* lived on muddy substrates down to a depth of 20,000 ft (6,000 m) or more.

REMARK As indicated by its usual (but incorrect) name, *Ventriculites infundibuliformis*, this species is generally funnel-shaped. It occurs commonly in the Late Cretaceous (Chalk) of Europe.

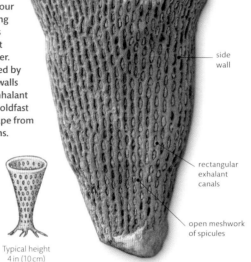

top

Rhizopoterion cribrosum (Phillips); Upper Chalk; Late Cretaceous; UK.

side wall

rectangular exhalant canals

open meshwork of spicules

Typical height 4 in (10 cm)

Range: Cretaceous	Distribution: Europe	Occurrence: ◉◉◉◉◉

Group: PHARENTRONIDA	Subgroup: LELAPIIDAE	Informal name: Calcisponge

Raphidonema

This genus belongs to the separate class of calcisponges, in which the skeleton is made of calcareous, rather than siliceous, spicules. *Raphidonema* is irregularly cup-shaped and very variable in size. Generally, the interior is smooth, while the outside is covered with rounded, knobby projections.

HABITAT This calcareous sponge lived in warm, shallow waters.

REMARK This is the most common sponge found in the famous Faringdon Sponge Gravels at Faringdon, Oxfordshire, in the United Kingdom.

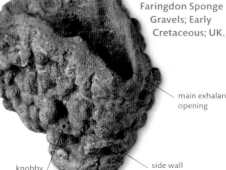

Raphidonema farringdonense (Sharpe); Faringdon Sponge Gravels; Early Cretaceous; UK.

main exhalant opening

side wall

knobby outside surface

Typical height 3¼ in (8 cm)

Range: Triassic–Cretaceous	Distribution: Europe	Occurrence: ◉◉◉◉◉

Group: ACTINOSTROMATIDA	Subgroup: ACINOSTROMATIDAE	Informal name: Stromatoporoid

Actinostroma

Actinostroma is a typical calcareous, reef-building, bloblike stromatoporoid. Although enigmatic, the genus is thought by some to be the ancestor of modern sclerosponges. When studied in thin section, it shows a fine structure of thin, concentric layers and radial pillars.

HABITAT This genus commonly lived on and formed reefs.
REMARK These organisms were important rock formers, particularly in the Silurian and Devonian.

concentric growth layers

irregular shape

Typical height
4¾ in (12 cm)

Actinostroma clathratum
Nicholson; Middle Devonian; UK.

Range: Cambrian–Early Carboniferous	Distribution: Worldwide	Occurrence: ◉◉◉◉◉

Group: ARCHAEOCYATHIDA	Subgroup: METACYATHIDAE	Informal name: Archaeocyathid

Metaldetes

Metaldetes is a widely distributed example of the group of marine animals known as archaeocyathids. Although sponges and archaeocyathids are unrelated, the latter have a similar calcareous (but nonspicular) skeleton. *Metaldetes* has basically a single- or double-walled cone perforated by numerous pores. The central cavity is empty, as in sponges. It is not known how the soft tissue was organized.

HABITAT These small creatures lived in reefs in warm, shallow seas.
REMARK Most archaeocyathids are found silicified in hard limestones and are known from the Early Cambrian of South Australia, Siberia, Sardinia, and Antarctica.

outer wall

Metaldetes taylori (Bedford); Early Cambrian; Australia.

septum (partition)

empty central cavity

Typical height
2 in (5 cm)

inner wall

Range: Cambrian	Distribution: Worldwide	Occurrence: ◉◉◉◉◉

BRYOZOANS

BRYOZOANS ARE colonial animals that resemble miniature corals but are more closely related to brachiopods. They range from the Ordovician to the present day. Most bryozoans live on the sea bed and secrete calcareous skeletons. Bryozoan colonies vary in shape—some are sheetlike encrustations on shells and stones, whereas others grow as small tree shapes or netlike fronds. Each colony consists of a few to thousands of connected individuals (zooids). Each zooid has a tubular or box-shaped skeleton.

Group: CYSTOPORATA	Subgroup: CONSTELLARIIDAE	Informal name: Sea mat

Constellaria

This bryozoan developed as a bushy colony with thick, often compressed branches. The branch surfaces are covered by distinctive star-shaped monticules (regularly shaped hummocks). Feeding zooids are located along the rays of the stars. The monticules of colonies probably formed chimneys for the expulsion of exhalant feeding currents.

HABITAT *Constellaria* lived on the sea bed.

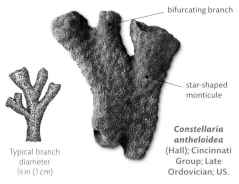

bifurcating branch

star-shaped monticule

Typical branch diameter ⅜ in (1 cm)

Constellaria antheloidea (Hall); Cincinnati Group; Late Ordovician; US.

Range: Ordovician–Silurian	Distribution: N. America, Europe, Asia	Occurrence: ◉

Group: CRYPTOSTOMATA	Subgroup: PTILODICTYIDAE	Informal name: Sea mat

Ptilodictya

This colony consists of a single branch, straight or gently curved, and diamond-shaped in cross-section, with a median wall butting box-shaped zooids from both sides. The zooids are rectangular in outline and arranged in longitudinal rows on the branch surface. At the pointed end of the branch is a conical structure that fitted into a socket on the encrusting base of the colony and permitted the branch to articulate.

HABITAT *Ptilodictya* lived in the sea with the encrusting base cemented to hard ground.
REMARK This specimen has a brachiopod lying next to it.

Ptilodictya lanceolata (Goldfuss); Wenlock Limestone; Late Silurian; UK.

associated brachiopod

Tapering branch end, articulated by a socket

Typical length 9 in (23 cm)

Range: Ordovician–Devonian	Distribution: Worldwide	Occurrence: ◉

| Group: CRYPTOSTOMATA | Subgroup: FENESTELLIDAE | Informal name: Lace corals |

Fenestella

This erect, netlike colony is formed of narrow, regularly spaced branches linked by dissepiments. The zooids are arranged in two rows along the branches, with apertures opening on one side only of the planar, folded, or conical colonies. The dissepiments lack zooids. Spinose outgrowths have been developed, especially near the colony base, and were sometimes barbed.

HABITAT *Fenestella* and similar reticulate bryozoans lived by filtering food from self-generated water currents, which flowed in one direction through the holes in the colony.

branches linked by dissepiments

limestone matrix

Typical colony height 2 in (5 cm)

Fenestella plebeia McCoy; Carboniferous Limestone; Early Carboniferous; UK.

oldest part of colony

rootlike spines

| Range: Silurian–Permian | Distribution: Worldwide | Occurrence: 🌑 |

| Group: CYCLOSTOMATA | Subgroup: CAVIDAE | Informal name: Sea mat |

Ceriocava

This is a bushy colony with bifurcating cylindrical branches up to 2 in (5.5 cm) in diameter. The branch surfaces are covered by a honeycomb of polygonal (typically hexagonal) zooidal apertures, many sealed by lids. Long, sack-shaped polymorphic zooids for larval brooding are occasionally found.

HABITAT This bryozoan formed low colonies on the sea floor, filtering food from the surrounding water.

cylindrical branches

soft matrix

Typical branch diameter ⅛ in (3 mm)

Ceriocava corymbosa Lamouroux; Bryozoan limestone; Middle Jurassic; France.

fractured end of branch

| Range: Jurassic–Cretaceous | Distribution: Europe | Occurrence: 🌑 |

Group: TUBULIPORTA	Subgroup: MULTISPARSIDAE	Informal name: Sea mat

Reptoclausa

This encrusting colony is characterized by having long ridges aligned parallel to the direction of its growth. Ridge flanks and crests are occupied by feeding zooids with subcircular apertures. Smaller, indistinct, nonfeeding, polymorphic zooids form the smooth-surfaced valleys between the ridges. These zooids lack apertures.

HABITAT These colonies were relatively robust and capable of living on rolling stones and shells in turbulent marine conditions.

ridges on colony surface

sediment trapped between ridges

sheetlike colony spreading over pebble

Typical colony diameter 1½ in (4 cm)

Reptoclausa hagenowi (Sharpe); Faringdon Sponge Gravel; Early Cretaceous; UK.

exposed surface of dark gray pebble

Range: Jurassic–Cretaceous	Distribution: Europe, Asia	Occurrence: ◉◉

Group: RETEPORIDAE	Subgroup: SERTELLIDAE	Informal name: Lace corals

Schizoretepora

This is a colony of complex, folded fronds pierced by oval fenestrules. Feeding zooids opened on one side of the fronds. They have apertures with sinuses. Defensive polymorphs (avicularia) are scattered over all branch surfaces. *Schizoretepora* is one of several similar genera that require careful microscopic study for accurate identification.

HABITAT *Schizoretepora* lives on the sea bed.

corrugated fronds

zooids opening on branch surface

Schizoretepora notopachys (Busk); Coralline Crag; Pliocene; UK.

Typical colony diameter 1½ in (4 cm)

perforations

cross-sections of broken branches

Range: Miocene–Recent	Distribution: Worldwide	Occurrence: ◉

Group: Not applicable	Subgroup: Not applicable	Informal name: Bryozoan limestone

Bryozoan limestone

Some limestones consist predominantly of fragments of the calcareous skeletons of bryozoan colonies. Many different bryozoan species may be present, often accompanied by broken mollusk shells and barnacles.

HABITAT Bryozoan limestones are especially common in the shallow-water deposits of the Cenozoic Era. Today, they can be found forming in subtropical to cold-water environments, but bryozoan limestones are rarer in the tropics, where coral limestones are found in great abundance.

echinoid fragment

pieces of reticulate bryozoans

branch fragments

Width of rock specimen 4½ in (11 cm)

bryozoan encrusting a shell

Bryozoan limestone; Miocene; Australia.

Range: Jurassic–Recent	Distribution: Worldwide	Occurrence: ●●●●●

Group: CHEILOSTOMATA	Subgroup: CLEIDOCHASMATIDAE	Informal name: Sea mat

Hippoporidra

The thick, multilayered colonies encrust gastropod shells. The feeding zooids have frontal walls pierced by marginal pores and apertures with a sinus. Atop the monticules are larger zooids, and scattered between the monticules are avicularia with pointed rostra.

HABITAT Recent *Hippoporidra* live symbiotically with hermit crabs, providing the crabs with excellently camouflaged homes. Although the crabs are not preserved as fossils, paleontologists believe the bryozoan colony form is diagnostic of their past presence.

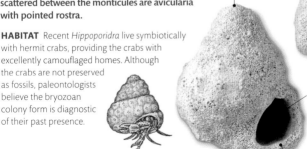

monticule

gastropod shell completely enveloped

apex

larger zooids on monticule

hermit crab aperture

Hippoporidra edax (Busk); James River Formation; Pleistocene; US.

Typical length ¾ in (2 cm)

outer lip

Range: Miocene–Recent	Distribution: Europe, N. America, Africa	Occurrence: ●●

WORMS

WORM IS A GENERAL NAME applied to representatives of various soft-bodied invertebrate groups. They include the platyhelminthes, nematodes, nemertines, acanthocephalans, annelids, and some hemichordates. Most are unknown or very rare in the fossil record. However, some members of the polychaeta group of segmented worms secrete a dwelling tube made of durable calcite, which is often found attached to fossil shells or pebbles. These are common fossils in certain Mesozoic and Cenozoic rocks.

Group: SABELLIDA	Subgroup: SERPULIDAE	Informal name: Serpulid worm

Proliserpula

Proliserpula ampullacea (J. Sowerby); Upper Chalk; Late Cretaceous; UK.

The tube of this worm is always attached to a hard substrate, often a fossil shell like the sea urchin shown here. It coils in an irregular spiral, with a single narrow ridge running along the midline. The aperture is circular, and the sides display prominent growth lines.

HABITAT This genus is found only in the deeper waters of the chalk seas.
REMARK There was great competition for hard surfaces to encrust, and other species have attached themselves to this shell.

Typical length 2 in (5 cm)

tube

circular opening for tentacles

spiral growth form

Range: Late Cretaceous	Distribution: Europe	Occurrence: ◉◉◉

Group: SABELLIDA	Subgroup: SERPULIDAE	Informal name: Serpulid worm

Rotularia

coiled tube

Rotularia bognoriensis (Mantell); London Clay; Early Eocene; UK.

The ridged living tube of *Rotularia* is tightly coiled in a flat spiral and slightly concave on one side and convex on the other. The last-formed part of the tube stands erect, like the tip of a fireman's hose.

HABITAT This form was free-living on sandy substrates in a shallow marine environment. The dense aggregation of individuals in this specimen was probably caused by storm waves.

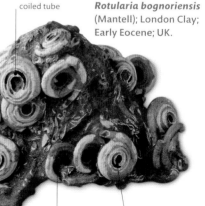

Typical diameter ⅝ in (1.5 cm)

concave face

convex face

Range: Eocene	Distribution: Europe	Occurrence: ◉◉◉◉◉

| Group: SABELLIDA | Subgroup: SERPULIDAE | Informal name: Serpulid worm |

Serpula

The tube of this serpulid worm is elongated and rounded in cross-section, tapering slowly to a fine point. It may be gently curved or sharply bent, with a relatively thin wall, ornamented by irregular growth lines. It has a rounded aperture.

Serpula indistincta (Fleming); Carboniferous Limestone; Early Carboniferous; UK.

HABITAT Although this genus was not cemented to the substrate, it was not actually mobile.

narrow part of tube, formed first

limestone matrix

rounded aperture

growth lines

Typical length 2 in (5 cm)

| Range: Paleozoic–Recent | Distribution: Worldwide | Occurrence: ◉ |

| Group: SABELLIDA | Subgroup: SERPULIDAE | Informal name: Serpulid worm |

Glomerula

The narrow, rounded tubes of *Glomerula* grew in an irregular, twisting fashion. Tubes from different individuals intertwined, so the whole group eventually resembled a badly rolled ball of broken string. In proportion to the thick wall of the tube, the aperture is actually quite narrow.

Glomerula plexus (J. Sowerby); Upper Chalk; Late Cretaceous; UK.

HABITAT A large number of individuals made up a single group, which lived unattached on the sea bed. This intertwined habit allowed it this freedom.
REMARK Like most serpulids, *Glomerula* fed by filtering particles of organic debris and plankton from the sea.

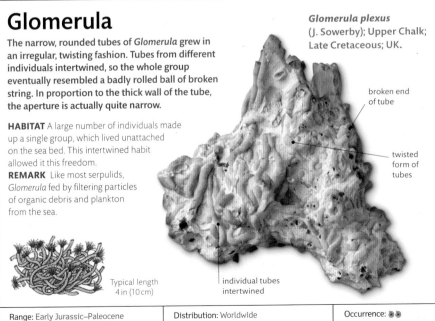

broken end of tube

twisted form of tubes

individual tubes intertwined

Typical length 4 in (10 cm)

| Range: Early Jurassic–Paleocene | Distribution: Worldwide | Occurrence: ◉◉ |

TRACE FOSSILS

TRACE FOSSILS ARE the remains of structures made by animals, preserved in sedimentary rock. They include tracks, trails, borings, and burrows. Because different organisms are able to make the same trace, trace fossils are classified by their shape rather than by their constructor. Ichnogenera and ichnospecies, from the Greek word *ichnos* (trace), are the usual classifications, but fossils may also be classified by their cause, such as feeding, crawling, dwelling, and so on, and hence fodinichnia, repichnia, dominichnia.

| Group: Unclassified | Subgroup: FODINICHNIA | Informal name: Trace fossil |

Chondrites

This small burrow superficially resembles a plant root, as it comprises a central tube, from the base of which numerous, subdividing branches radiate out. The animal most likely to have constructed these burrows is a nematode worm (roundworm).

HABITAT *Chondrites* is found in marine sediments and is especially common in sediment formed in conditions of reduced oxygen.

Chondrites sp.; Late Cretaceous; Spain.

mudstone matrix

branching, sediment-filled burrows

single central tube

Variable length

| Range: Triassic–Recent | Distribution: Worldwide | Occurrence: ◉◉◉◉◉ |

| Group: Unclassified | Subgroup: REPICHNIA | Informal name: Trace fossil |

Cruziana

This trace has a two-lobed structure with a central groove—the hardened infill of a double groove excavated on the sea floor. The lobes are covered with scratch marks made by the legs of the excavating organism, usually a trilobite.

HABITAT *Cruziana* is most common in marine sediments from the Paleozoic era.

sandstone preservation

infill of double groove

Variable length

Cruziana sp.; Nubian sandstone; Cambrian; Egypt.

scratch mark made by leg

| Range: Cambrian–Triassic | Distribution: Worldwide | Occurrence: ◉◉◉ |

PROBLEMATICA

SOME FOSSILS CANNOT be placed into major groups of animals or plants with any certainty. Often this is because they represent only a small, obscure part of an organism, but there are others, from Ediacaran sediments, that have completely preserved forms. These have been variously classified as worms, soft corals, jellyfish, and so on. However, the true identity of these ancient fossils remains a subject of controversy—some may even be trace fossils—and they are best treated as Problematica.

Group: Unclassified	Subgroup: Unclassified	Informal name: Trace fossil

Mawsonites

This large fossil has a circular outline, broken into irregular lobes. The surface carries irregular lobe- or scale-shaped bosses arranged concentrically, with a central, buttonlike swelling. Originally interpreted as a jellyfish, it is now thought this form may have been a radiating burrow system and thus a trace fossil.

HABITAT This structure could have been made by an organism hunting through silt for food.

Mawsonites spriggi
Glaessner & Wade;
Ediacara sandstone;
Ediacaran; Australia.

irregular, lobed margin

concentric lobes

central, buttonlike protuberance

Typical length
5 in (13 cm)

Range: Ediacaran	Distribution: Australia	Occurrence: ◉

Group: PENNATULACEA	Subgroup: CHARNIIDAE	Informal name: Sea pen

Charniodiscus

central axis

This is a feather-shaped fossil attached to a basal disk (not present on this specimen). Closely spaced alternate branches diverge from a central axis, with each branch subdivided by about 15 transverse grooves.

HABITAT This animal lived attached to the sea floor, feeding by filtering nutritious particles from the water.

REMARK The affinities of *Charniodiscus* remain uncertain, but most paleontologists consider it to be a type of soft coral. One theory suggests it is a representative of a distinct group of organisms, the Vendozoa.

Charniodiscus masoni (Ford);
Woodhouse Beds;
Ediacaran; UK.

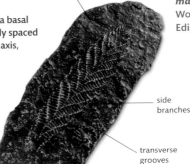

side branches

transverse grooves

Typical length
8 in (20 cm)

Range: Ediacaran	Distribution: Australia, Europe	Occurrence: ◉

| Group: Unclassified | Subgroup: Unclassified | Informal name: Unclassified |

Spriggina

The body of *Spriggina* is elongated, tapering to a point at the rear, similar to a tail. The broader front end is rounded and consists of a narrow, crescent-shaped arc. The rest of the body is divided along its length by a central line, and transversely into short, broadly V-shaped segments.

Spriggina floundersi Glaessner; Ediacara sandstone; Ediacaran; Australia.

HABITAT *Spriggina* lived in a shallow, sandy marine habitat.

REMARK This form was first thought to be a worm, then an early arthropod or part of a new group, neither plant nor animal.

pointed "tail" end

V-shaped segments

crescent-shaped "head"

Typical length 2¾ in (7 cm)

| Range: Ediacaran | Distribution: Australia, Africa, Russia | Occurrence: ◉ |

| Group: CONULARIIDA | Subgroup: CONULARIIDAE | Informal name: Conularid |

Paraconularia

This shell is conical and four-sided but rhombic in cross-section. Each of the four sides carries upwardly arched, closely spaced growth lines, and at each of the four corners is a shallow V-shaped groove. The shell itself is thin and made of phosphatic and proteinaceous material.

Paraconularia derwentensis (Johnston); Early Permian; Australia.

tentacled end in the living animal

HABITAT This species lived in shallow to deep marine waters, attached to the sea bed by a small basal disk.

REMARK Members of this family have been variously placed in the mollusks, in a new group related to the wormlike phoronids, or most commonly with the jellyfish because of the supposed tentacles sometimes preserved. The tentacles may have been used to catch small marine animals, stinging in the same way as sea anemones.

growth lines

broken base

Typical length 4 in (10 cm)

| Range: Early Permian | Distribution: Australia | Occurrence: ◉◉ |

GRAPTOLITES

GRAPTOLITES ARE AN extinct group of colonial organisms that lived from the Cambrian to the Carboniferous. Their remains can be mistaken for fossil plants, as they sometimes resemble fossilized twigs. Each colony had one or more branches (stipes), originating from an individual (sicula). Each subsequent individual was housed within a tubular structure (theca). Geologists find graptolites especially useful in dating rocks.

Group: GRAPTOLOIDEA	Subgroup: RETIOLITIDAE	Informal name: Graptolite

Retiolites

The thecae of *Retiolites* are arranged in two series lying back to back, forming a single stipe, which gradually expands in width away from the proximal end. The organic-walled skeleton of *Retiolites* had a characteristic open network, which presumably made it lighter in the water.

HABITAT *Retiolites* probably had a free-floating existence.
REMARK Free-floating graptolites, such as *Retiolites*, make good zonal fossils.

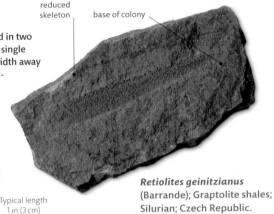
reduced skeleton   base of colony

Typical length 1 in (3 cm)

Retiolites geinitzianus (Barrande); Graptolite shales; Silurian; Czech Republic.

Range: Silurian	Distribution: Worldwide	Occurrence: ●●●●●

Group: GRAPTOLOIDEA	Subgroup: DICHOGRAPTIDAE	Informal name: Graptolite

Phyllograptus

Phyllograptus typus Hall; Levis shale; Early Ordovician; Canada.

This colony comprised four series of thecae arranged back-to-back in a crosslike formation. Usually, due to compression, only a pair of thecae are clearly visible on shale specimens. The thecae are long and curved with small apertural lips. The generic name, *Phyllograptus*, refers to the leaflike appearance of these flattened colonies.

HABITAT *Phyllograptus* floated freely in the open ocean.

Typical length 1⅜ in (3.5 cm)   shale matrix

Range: Early–Middle Ordovician	Distribution: Worldwide	Occurrence: ●●●●●

Group: GRAPTOLOIDEA	Subgroup: DICHOGRAPTIDAE	Informal name: Graptolite

Expansograptus

These were small to large graptolites with approximately 20 to several hundred thecae. Two stipes diverge from the initial sicula, which is distinguished by its dorsally projecting nema (an extension of the initial sicula) at the point where the thecae of the two stipes diverge. The thecae are usually of relatively simple type, tubular or with a small apertural lip. The different species are distinguished by the number of thecae per centimeter and the width and shape of the stipes.

HABITAT *Expansograptus* floated in open seas.

Expansograptus cf. *nitidus* (Hall); Mytton Formation; Early Ordovician; UK.

two extended stipes

sicula

Typical length 2 in (5 cm)

Range: Early–Middle Ordovician	Distribution: Worldwide	Occurrence: ◉◉◉◉◉

Group: GRAPTOLOIDEA	Subgroup: DICHOGRAPTIDAE	Informal name: Graptolite

Loganograptus

These were large graptolites; the rapid initial dichotomies near the sicula produced a colony with 16 stipes. These stipes were evenly separated in life but frequently bent in fossils. The thecae are simple tubes—about one per millimeter.

HABITAT This genus lived in the deeper parts of the Ordovician ocean.

Loganograptus logani (Hall); Skiddaw Group; Early Ordovician; UK.

thecae slowly increase in size along stipes

Typical length 8 in (20 cm)

Range: Early Ordovician	Distribution: Worldwide	Occurrence: ◉◉◉◉◉

Group: GRAPTOLOIDEA	Subgroup: DICHOGRAPTIDAE	Informal name: Graptolite

Tetragraptus

This graptolite had four stipes of variable form—they may have been reclined or outstretched horizontally. Some species had a weblike structure connecting the four stipes. The thecae are usually of simple, tubular form, and may be inclined at high or low angles. Stipes vary in width from ⅕₅ to ⅖ in (1–10 mm).

HABITAT *Tetragraptus* floated widely in Early Ordovician seas.

fissile shale

partially visible sicula

tubular thecae

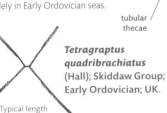

Tetragraptus quadribrachiatus (Hall); Skiddaw Group; Early Ordovician; UK.

Typical length 4 in (10 cm)

Range: Early Ordovician	Distribution: Worldwide	Occurrence: ●●●●●

Group: GRAPTOLOIDEA	Subgroup: DIPLOGRAPTIDAE	Informal name: Graptolite

Orthograptus

Orthograptus was a biserial graptolite, which means that it had two series of thecae lying back-to-back, producing a robust, double-edged colony. The proximal growth of the colony is a crucial identification feature. *Orthograptus* had a strongly developed spine. The thecal apertures were moderately straight. The variations in colony size, width, and thecal form determine the species.

HABITAT This graptolite is possibly epiplanktonic, and floated in very shallow seas.

mudstone

carbonized periderm

robust colony

Orthograptus intermedius (Elles); Bifidus Beds; Early Ordovician; UK.

Typical length 3 in (8 cm)

Range: Middle–Late Ordovician	Distribution: Worldwide	Occurrence: ●●●●●

| Group: GRAPTOLOIDEA | Subgroup: MONOGRAPTIDAE | Informal name: Graptolite |

Monograptus

Graptolites of this genus had a single stipe and a highly variable form; they could be completely straight, gently curved, or completely spiral. The form of the thecae is vital to identification. In this genus, there was a tendency for thecae to become isolated from one another. Some evolved very rapidly, so species of this genus are easily recognizable and stratigraphically useful.

HABITAT It floated in open seas.

Monograptus convolutus
Hisinger; Silurian Flags; Silurian; UK.

Typical diameter 8 in (20 cm)

isolated distal thecae

triangulate thecae

| Range: Silurian–Early Devonian | Distribution: Worldwide | Occurrence: ◉◉◉◉◉ |

| Group: GRAPTOLOIDEA | Subgroup: DICHOGRAPTIDAE | Informal name: Graptolite |

Didymograptus

These graptolites formed a shape similar to a tuning fork. The colonies were often robust, ranging in size from ⅜ to 4 in (1–10 cm). The thecae are long with simple tubes. The stipes expand in width from the proximal end.

HABITAT This genus floated in northern oceans.

Didymograptus murchisoni
(Beck); Llanvirn Series; Early Ordovician; UK.

dark shale

siculae at pointed ends

mineralized periderm

Typical length 4 in (10 cm)

| Range: Early–Middle Ordovician | Distribution: Worldwide | Occurrence: ◉◉◉◉◉ |

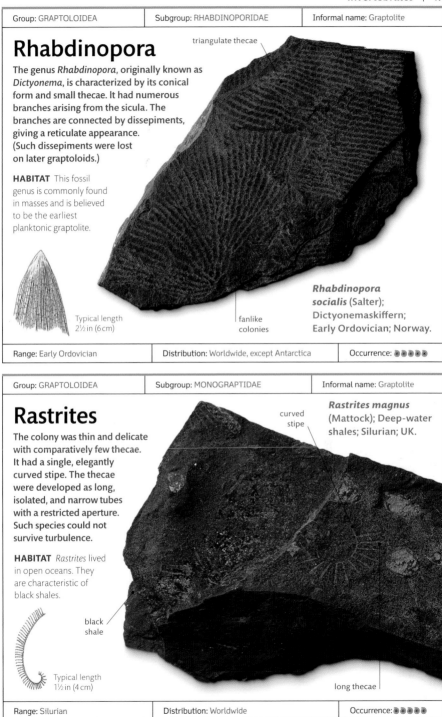

| Group: GRAPTOLOIDEA | Subgroup: RHABDINOPORIDAE | Informal name: Graptolite |

Rhabdinopora

triangulate thecae

The genus *Rhabdinopora*, originally known as *Dictyonema*, is characterized by its conical form and small thecae. It had numerous branches arising from the sicula. The branches are connected by dissepiments, giving a reticulate appearance. (Such dissepiments were lost on later graptoloids.)

HABITAT This fossil genus is commonly found in masses and is believed to be the earliest planktonic graptolite.

Typical length
2½ in (6 cm)

fanlike
colonies

Rhabdinopora socialis (Salter); Dictyonemaskiffern; Early Ordovician; Norway.

| Range: Early Ordovician | Distribution: Worldwide, except Antarctica | Occurrence: ⊛⊛⊛⊛⊛ |

| Group: GRAPTOLOIDEA | Subgroup: MONOGRAPTIDAE | Informal name: Graptolite |

Rastrites

Rastrites magnus (Mattock); Deep-water shales; Silurian; UK.

curved stipe

The colony was thin and delicate with comparatively few thecae. It had a single, elegantly curved stipe. The thecae were developed as long, isolated, and narrow tubes with a restricted aperture. Such species could not survive turbulence.

HABITAT *Rastrites* lived in open oceans. They are characteristic of black shales.

black shale

Typical length
1½ in (4 cm)

long thecae

| Range: Silurian | Distribution: Worldwide | Occurrence: ⊛⊛⊛⊛⊛ |

CORALS

CORALS ARE MARINE ANIMALS with a saclike body (polyp), mouth, tentacles, and skeleton. The polyp occupies a circular, polygonal, or elongate cavity (calice), which is generally surrounded by a wall. Calices are usually divided by starlike arrangements of plates (septa), which sometimes have a central structure. The calice wall, if present, forms horn- or tubelike features (corallites), often divided by simple transverse partitions (tabulae), by series of plates (dissepiments), or both. Polyps may divide to form colonies, with corallites sometimes joined by intercorallite structures (coenosteum). There are three main coral groups, two extinct (Rugosa, Tabulata) and one extant (Scleractinia).

| Group: CYSTIPHYLLIDA | Subgroup: GONIOPHYLLIDAE | Informal name: Rugose coral |

Goniophyllum

This coral is usually solitary. It has a square to quadrilateral transverse section and a lidlike structure of four plates (sometimes missing). The calice is deep and the short septa bear distinctive flanges. There are numerous internal dissepimental plates.

HABITAT *Goniophyllum* lived in shallow-water limestones and muds.

Typical calice diameter ⅝ in (1.5 cm)

angular corallite

short septa

Goniophyllum pyramidale Hisinger; Silurian; Sweden.

| Range: Early–Middle Silurian | Distribution: Europe, N. America | Occurrence: ◉◉ |

| Group: STAURIIDA | Subgroup: ZAPHRENTIDAE | Informal name: Rugose coral |

Heliophyllum

A solitary, sometimes colonial coral with open branched or adjoined corallites. The calice is shallow with long septa, sometimes reaching the center of the corallite. Dissepiments are numerous, small, well-rounded, and confined to the outer zone of corallite. The central zone is occupied by flat or slightly curved tabulae. The external walls are thin with fine growth ridges.

HABITAT *Heliophyllum* inhabited shallow water.

Heliophyllum sp.; Hamilton Formation; Middle Devonian; Canada.

horn-shaped corallite

rejuvenating calice

fine growth ridges

Typical calice diameter ¼ in (2 cm)

| Range: Devonian | Distribution: Worldwide, except Asia | Occurrence: ◉◉ |

Group: STAURIIDA	Subgroup: LITHOSTROTIONIDAE	Informal name: Rugose coral

Siphonodendron

This is a colonial coral with branches separate from each other, with one corallite center to each branch. The colonies are bushy, with corallites radiating outward, often becoming more or less parallel. The axial structure is a distinctive flattened rod.

Polished specimen

Siphonodendron junceum Fleming; Carboniferous Limestone; Early Carboniferous; UK.

HABITAT *Siphonodendron* is found in shallow-water limestones and muds.
REMARK The name *Siphonodendron* is now used for forms previously called *Lithostrotion* that have branched colonies. *Lithostrotion* is still used for forms with closely adjoined corallites.

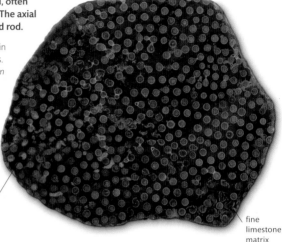

transverse sections of corallites

fine limestone matrix

Typical calice diameter 1 in (2.5 cm)

Range: Carboniferous	Distribution: Worldwide	Occurrence: ⊚

Group: FAVOSITIDA	Subgroup: FAVOSITIDAE	Informal name: Tabulate coral

Favosites

Favosites has flattened to hemispherical colonies. The corallites are polygonal to slightly rounded; closely adjoined, giving colonies a honeycomblike appearance; and very variable in size. The calices are concave, with fewer than 12 short, equal-length septa. The walls are thin, perforated by small pores in four or fewer longitudinal rows. Tabulae are numerous, more or less flat, and horizontal.

weathered colony surface

polygonal to slightly rounded corallites

HABITAT *Favosites* lived in shallow water, including reefs and calcareous shales.

Typical calice diameter 1/16 in (2 mm)

Favosites sp.; Wenlock Limestone; Late Silurian; UK.

concave calices

Range: Ordovician–Devonian	Distribution: Worldwide	Occurrence: ⊚⊚⊚⊚⊚

| Group: STAURIIDA | Subgroup: AXOPHYLLIDAE | Informal name: Rugose coral |

Actinocyathus

This colonial coral has flattened to rounded colonies of closely adjoined polygonal corallites of different sizes in a coarsely honeycomblike pattern. The calices are concave, with a central boss. The septa are thin, often alternating long and short. The axial boss is complex, consisting of a steeply sloping series of small blistery plates arranged in conical form around a small central plate and also intersected by septal plates. The dissepiments are in series, forming a wide outer zone within the corallites, distinctively very large and irregular. In transverse section, the septa appear to be largely absent from the dissepimental zone but actually form fine ridges running across upper surfaces of dissepiments. The tabulae are flat or concave. The corallites have thin external walls with fine growth ridges, but these are only visible in well-preserved specimens when broken along the corallite junctions.

HABITAT *Actinocyathus* lived in shallow-water limestones and mud.
REMARK The name *Actinocyathus* is now used for forms previously called *Lonsdaleia* with closely adjoined corallites. *Lonsdaleia* is still used for openly branched forms.

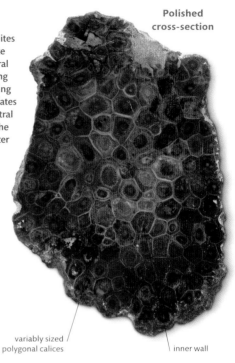

Polished cross-section

variably sized polygonal calices

inner wall

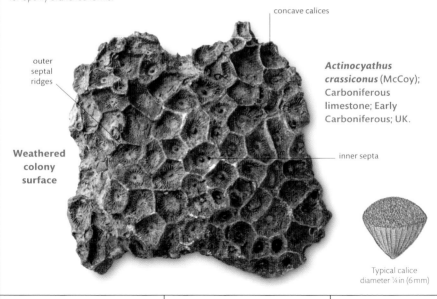

concave calices

outer septal ridges

Weathered colony surface

inner septa

Actinocyathus crassiconus (McCoy); Carboniferous limestone; Early Carboniferous; UK.

Typical calice diameter ¼ in (6 mm)

| Range: Carboniferous | Distribution: Europe, Asia | Occurrence: ⦿⦿⦿⦿⦿⦿ |

Group: FAVOSITIDA	Subgroup: HALYSITIDAE	Informal name: Chain coral

Halysites

This colonial coral has its corallites arranged in single series in "pan-pipe" formation, with a minute tube between each corallite. In transverse section, they appear chainlike. The chains are straight to curved, dividing and rejoining each other, enclosing cell-like spaces between them. The individual corallites are rounded to elliptical. Septa are absent or present as a few inconspicuous spines. The walls are thick and the tabulae are numerous, mostly flat and horizontal.

HABITAT *Halysites* inhabited warm, shallow waters, including reefs.

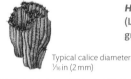

Halysites catenularius (Linnaeus); Niagara group; Silurian; US.

Typical calice diameter 1⁄16 in (2 mm)

corallites

Colony from above

sediment in spaces between chains

Range: Middle Ordovician–Late Silurian	Distribution: Worldwide	Occurrence: ◉◉◉◉◉

Group: STAURIIDA	Subgroup: ARACHNOPHYLLIDAE	Informal name: Rugose coral

Arachnophyllum

This colonial coral has a low, shallow-domed coral head composed of polygonal calices. Each has a central columella resembling a spider's web. This is surrounded by a ring of alternating short and long septae and a broad outer dissepimental zone.

HABITAT *Arachnophyllum* inhabited warm, shallow-water limestones and muds.
REMARK Pieces of coralline limestone are gathered from the western Sahara to use as an ornamental stone. Glacially rolled pebbles of *Hexagoniaria*, a very similar coral called Petoskey stones, can be found on the shores of Lake Michigan.

Typical calice diameter 1⁄2 in (12 cm)

dissepiments   columella

Polished heart-shaped ornament

septa

Arachnophyllum pentagonum (Goldfuss); Devonian limestone; Morocco.

calices

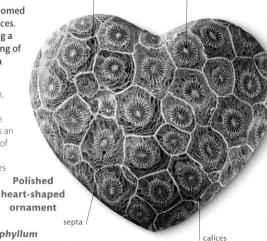

Range: Silurian–Devonian	Distribution: Africa, North America	Occurrence: ◉◉◉

Group: SCLERACTINIA	Subgroup: MONLIVALTIIDAE	Informal name: Colonial coral

Thecosmilia

Thecosmilia is a colonial coral with branches separate from each other. It has one, occasionally two, large corallite centers to each branch. The colonies are bushy with corallites radiating outward, often becoming more or less parallel. The calices are deeply concave. The septa are numerous, variable in size in transverse view, with successively shorter septa. The outsides of the branches bear toothed longitudinal ridges, which are continuous with the septa.

HABITAT *Thecosmilia* inhabited warm, shallow waters and reefs.

Typical calice diameter ½ in (1.2 cm)

rejuvenating corallite

sediment between corallites

Colony from above

Thecosmilia trichotoma (Goldfuss); "Corallian"; Late Jurassic; Germany.

septa arranged according to size

Range: Jurassic–Cretaceous	Distribution: Worldwide	Occurrence: ◉◉◉◉◉

Group: SCLERACTINIA	Subgroup: FAVIIDAE	Informal name: Brain coral

Colpophyllia

This is a dome-shaped or flattened, colonial coral. The corallites are closely adjacent, with shared walls. In transverse view, they are very elongate, variable in length, with the longer corallites meandering. The calices are concave, valleylike, and quite deep, separated by acute, ridgelike walls. The septa slope down into "valleys." There are numerous blistery dissepiments within the corallites.

HABITAT *Colpophyllia* lives in warm, shallow seas and reefs.

Typical calice diameter ⅜ in (1 cm)

sharp, corallite walls

Colony surface

Colpophyllia stellata (Catullo); Castelgomberto Limestone; Late Oligocene; Italy.

meandering, calicinal "valleys"

Range: Eocene–Recent	Distribution: Europe, N. & S. America	Occurrence: ◉◉◉◉

Group: SCLERACTINIA	Subgroup: FAVIIDAE	Informal name: Solitary coral

Trachyphyllia

Trachyphyllia is tapered to a sharp point; in transverse section, it is elliptical to elongate. The calices are gently concave to deep. There are numerous septa, variable in size; in transverse section, successively shorter septa are arranged in sets between those of the next longer size. The septal edges bear numerous fine teeth. The external wall is thin, sometimes discontinuous, covered with fine growth ridges. The internal wall is supported by arched dissepiments.

elongate calice

external wall with fine growth ridges

HABITAT
Trachyphyllia lives unattached on the sea floor in sheltered areas of flat soft sand in and around reefs.

Typical calice head
1½ in (4 cm)

Trachyphyllia chipolana Vaughan; Chipola Formation; Miocene; US.

Range: Miocene–Recent	Distribution: Worldwide	Occurrence: ◉

Group: SCLERACTINIA	Subgroup: RHIZANGIIDAE	Informal name: Horn coral

Septastraea

This is a colonial coral with a highly variable morphology. The colonies form irregular nodular masses, sometimes also branched, with numerous corallites within each branch. The corallites are closely adjoining, usually with shared walls, giving them a honeycomb appearance. There are usually 12 septa, all reaching the corallite center. The interiors of the corallites have a few thin dissepiments.

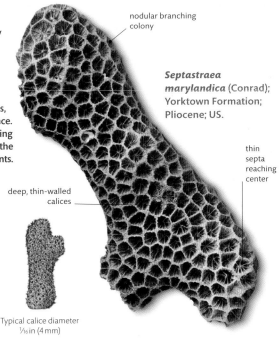

nodular branching colony

Septastraea marylandica (Conrad); Yorktown Formation; Pliocene; US.

thin septa reaching center

deep, thin-walled calices

HABITAT This horn coral inhabited warm, shallow-water reefs.
REMARK This coral is sometimes found encrusting marine snail shells, which were probably inhabited by hermit crabs. The coral continued growth outward from the shell, extending the space that could be occupied by the crab. There was probably a symbiotic association between the crab and the coral.

Typical calice diameter
⅙ in (4 mm)

Range: Miocene–Pleistocene	Distribution: N. America, Europe	Occurrence: ◉◉◉◉

TRILOBITES

ALTHOUGH THEY ARE now extinct, these arthropods flourished in the sea from the Cambrian through to the Permian. They ranged in length from ⅟₂₅ inch to 39 inches (1 mm to 1 m). The name, trilobite, derives from their division into three longitudinal lobes: a slightly raised central lobe (the axis), with two flatter pleural lobes on either side. They were also divided into a head shield (cephalon), a thorax of up to 30 segments, and a tail shield (pygidium). The axial region of the head shield (glabella) had cheeks on either side and often well-developed eyes. Each segment of the thorax had limbs, but these are rarely preserved. Trilobites could roll up (enroll) their external skeletons, probably for defense.

Group: AGNOSTIDA	Subgroup: AGNOSTIDAE	Informal name: Trilobite

Acadagnostus

This was a small, blind trilobite, with only two thoracic segments, but it was still able to roll up tightly. It has a furrow dividing the area in front of the glabella and a similar furrow behind the axis of the tail shield. The head and tail are the same size.

HABITAT This genus lived in deep water.

cephalon

glabella

tail shield

Typical length
⅓ in (8 mm)

Acadagnostus exaratus (Grönwall); Menevian Series; Middle Cambrian; UK.

Range: Middle Cambrian	Distribution: N.E. America, Europe, Australia	Occurrence: ◉◉◉◉◉

Group: REDLICHIIDA	Subgroup: OLENELLIDAE	Informal name: Trilobite

Olenellus

This early trilobite has a small tail, large head, large eyes, and numerous spined thoracic segments, one of which is conspicuously wider than the rest. The thorax itself tapers gradually backward. Dorsal facial sutures—a feature of most other trilobites—had not yet developed in *Olenellus*. The head shield is surrounded by a narrow rim, which continues into the genal spines. The cuticle of the external skeleton is thin.

HABITAT This not-very-active swimmer lived near the sea bed.

long, crescent-shaped eyes

thin, needlelike genal spines

long pleural spine

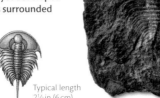

Typical length
2½ in (6 cm)

Olenellus thomsoni Hall; Olenellian Series; Early Cambrian; US.

Range: Early Cambrian	Distribution: Scotland, N. America	Occurrence: ◉◉◉◉

| Group: PTYCHOPARIIDA | Subgroup: OLENIDAE | Informal name: Trilobite |

Triarthrus

The external skeleton of this trilobite is more than twice as long as it is wide. It has a large head shield and a small tail shield. There are no genal spines, but two pairs of deep furrows are visible on the glabella.

HABITAT *Triarthrus* lived on or near the sea bed.
REMARK This specimen is one of the rare examples in which the limbs of the trilobite are preserved—the traces of the walking legs may be clearly seen. These were preserved because they were covered at an early stage of burial by a film of iron pyrites, which remains even after the soft tissue has decayed.

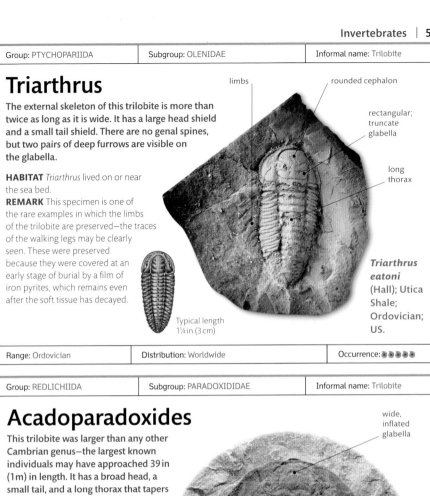

limbs

rounded cephalon

rectangular; truncate glabella

long thorax

Triarthrus eatoni (Hall); Utica Shale; Ordovician; US.

Typical length 1¼ in (3 cm)

| Range: Ordovician | Distribution: Worldwide | Occurrence: ◉◉◉◉◉ |

| Group: REDLICHIIDA | Subgroup: PARADOXIDIDAE | Informal name: Trilobite |

Acadoparadoxides

This trilobite was larger than any other Cambrian genus—the largest known individuals may have approached 39 in (1 m) in length. It has a broad head, a small tail, and a long thorax that tapers gradually and to which numerous segments were added as it grew. The eyes are well developed, the frontal lobe of the glabella inflated, and the tips of the lateral cheeks extended into long spines.

HABITAT This trilobite lived in shelf muds.

wide, inflated glabella

genal spines

Acadoparadoxides briareus (Geyer); Middle Cambrian; Morocco.

Typical length 8 in (20 cm)

long thorax

small tail shield

| Range: Middle Cambrian | Distribution: N. Africa | Occurrence: ◉◉◉◉ |

Group: REDLICHIIDA	Subgroup: XYSTRIDURIDAE	Informal name: Trilobite

Xystridura

This trilobite is broadly oval in outline. Its head section is about twice as wide as it is long, with a well-furrowed glabella and large eyes. The genal angle has extended to form short genal spines. The thorax is divided into 13 segments, with the axis being much narrower than the deeply furrowed pleural lobes. The tail shield is also furrowed, and spines are present around the posterior margin.

HABITAT *Xystridura* lived on or near the sediment of the sea bed.

Xystridura saint-smithii (Chapman); Middle Cambrian; Australia.

large eyes

thorax with short spines

moderately large tail shield

Typical length 2½ in (6 cm)

Range: Middle Cambrian	Distribution: Australia	Occurrence: ◉◉

Group: ASPHATIDA	Subgroup: TRINUCLEIDAE	Informal name: Trilobite

Declivolithus

This trilobite is wide in relation to its length. The convex head section is surrounded by a pitted fringe, with the irregular arrangement of the small pits being characteristic of the genus. The thorax is made up of six narrow segments, followed by a furrowed, triangular tail shield. The narrow axis reaches almost to the margin.

HABITAT *Declivolithus* burrowed just below the surface of the sea bed.
REMARK This genus contains the largest species of trinucleid trilobites.

pitted cephalic fringe

large cephalon

short thorax

furrowed tail shield   **Internal mold**

Declivolithus titan Fortey and Edgecombe; Ordovician; Morocco.

Typical length 1¼ in (3 cm)

Range: Middle–Late Ordovician	Distribution: Europe, N. Africa	Occurrence: ◉◉◉

| Group: PHACOPIDA | Subgroup: CHEIRURIDAE | Informal name: Trilobite |

Ktenoura

This trilobite has an elongate external skeleton, with a thorax made up of 11 segments. Facial sutures are of the proparian type, cutting into the side margin of the head. The cheeks and pleurae are narrow, with the eyes situated anteriorly and close to the glabella. The glabella itself is convex and expands gently forward in width. The complex boss-and-socket articulation between segments allowed it to roll up tightly. It has a moderately sized tail shield, with three pairs of long, marginal spines that curve backward.

HABITAT This trilobite was a typical inhabitant of shelf seas.

Typical length
2 in (5 cm)

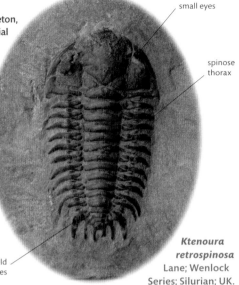

small eyes

spinose thorax

tail shield with long spines

Ktenoura retrospinosa Lane; Wenlock Series; Silurian; UK.

| Range: Late Ordovician–Silurian | Distribution: Worldwide | Occurrence: ◉◉ |

| Group: PHACOPIDA | Subgroup: CHEIRURIDAE | Informal name: Trilobite |

Sphaerexochus

The external skeleton of this trilobite is convex and elongate, with a thick, calcite cuticle. Its raised central lobe is especially convex, and the inflated glabella occupies the bulk of the head width. The genal spines are reduced. Its thorax has 11 segments, with spiny pleural tips, curved downward. The short tail shield is also spiny.

HABITAT
Sphaerexochus probably lived in fairly shallow water on and around coral reefs.
REMARK This genus is usually found in Late Ordovician and Silurian limestones.

Typical length
1¼ in (3 cm)

glabella

swollen glabellar lobes

thorax

small tail shield

Sphaerexochus mirus Beyrich; Wenlock Limestone; Silurian; UK.

| Range: Ordovician–Silurian | Distribution: Worldwide | Occurrence: ◉◉◉ |

Group: LICHIDA	Subgroup: SELENOPELTIDAE	Informal name: Trilobite

Selenopeltis

This trilobite has a wide external skeleton, with the head and tail sections short in relation to the thorax. The wide glabella has complex furrows and an inflated central lobe. The genal spines are very long and the pleural tips of each thoracic segment are prolonged into great spines, which project backward. One similar pair of spines is found on the tail shield.

HABITAT *Selenopeltis* is thought to have floated freely in ocean waters.

wide glabella

long spinose tips

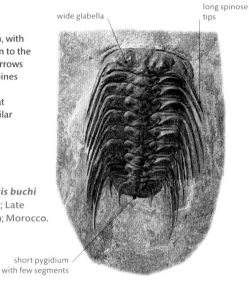

Selenopeltis buchi
(Barrande); Late
Ordovician; Morocco.

Typical length
5½ in (14 cm)

short pygidium
with few segments

Range: Ordovician	Distribution: Europe, N. Africa	Occurrence: ◉◉

Group: PHACOPIDA	Subgroup: PHACOPIDAE	Informal name: Trilobite

Eldredgeops

Enrolled specimen

large eye

This trilobite has a convex glabella that expands forward. The large eyes have fewer lenses than usual, which may have improved optics. The thorax has 12 segments, equipped with facets that facilitate enrollment. The deeply furrowed tail shield is smaller than its head. A sculpture of tubercles ornaments the thick cuticle.

HABITAT *Eldredgeops* inhabited shallow water in warm seas.

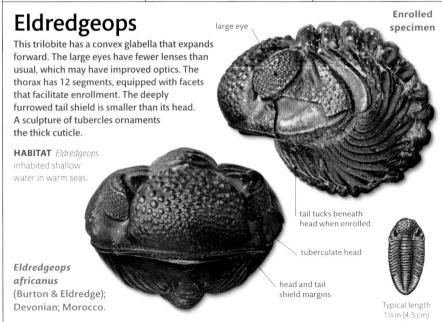

tail tucks beneath
head when enrolled

tuberculate head

*Eldredgeops
africanus*
(Burton & Eldredge);
Devonian; Morocco.

head and tail
shield margins

Typical length
1¾ in (4.5 cm)

Range: Devonian	Distribution: Worldwide	Occurrence: ◉◉◉◉◉

Group: PHACOPIDA	Subgroup: CALYMENIDAE	Informal name: Trilobite

Calymene

Trilobites of this genus have a convex external skeleton, with pleural tips turned down steeply, and a tail shield smaller than the head. The smallish eyes are placed close to the deeply furrowed glabella.

HABITAT This trilobite was a sluggish swimmer, so it probably walked on the sea bed.
REMARK In the 19th century, miners called it the "Dudley Locust."

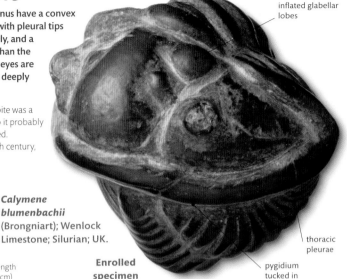

inflated glabellar lobes

thoracic pleurae

pygidium tucked in

Calymene blumenbachii (Brongniart); Wenlock Limestone; Silurian; UK.

Typical length 2¾ in (7 cm)

Enrolled specimen

Range: Late Ordovician–Silurian	Distribution: Worldwide	Occurrence: ◉◉◉◉

Group: LICHIDA	Subgroup: ODONTOPLEURIDAE	Informal name: Trilobite

Leonaspis

Leonaspis has a wide head drawn out to the side and rear into exceptionally stout genal spines and with small, slightly elevated eyes. The 10 thoracic segments have wide pleurae extended into robust spines and straight pleural furrows. The tail shield is also very spiny.

HABITAT A bottom dweller, this genus used its spines for protection.
REMARK This type of trilobite had a long stratigraphic range.

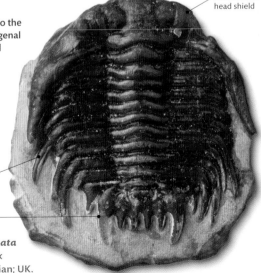

downturned head shield

very stout pleural spines

paired tail-shield spines

Typical length ⅝ in (1.5 cm)

Leonaspis coronata (Salter); Wenlock Limestone; Silurian; UK.

Range: Silurian–Devonian	Distribution: Worldwide	Occurrence: ◉◉◉

| Group: PHACOPIDA | Subgroup: DALMANITIDAE | Informal name: Trilobite |

Huntoniatonia

The exoskeleton of *Huntoniatonia* is two-thirds as wide as it is long. At the front of the cephalic shield is a short but robust "snout," typical of the genus, while at the back of the head there are exceptionally stout genal spines. The eyes are very large, strongly curved, and slightly elevated, with large lenses. The glabella extends forward and has three pairs of side furrows, the front pair of which slopes backward. The large, segmented tail shield is triangular in outline, with numerous segments and a pointed tip.

HABITAT This was an active, bottom-dwelling trilobite with a well-developed visual system. It may have had predatory habits.

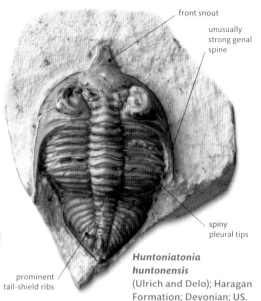

front snout

unusually strong genal spine

spiny pleural tips

prominent tail-shield ribs

Huntoniatonia huntonensis (Ulrich and Delo); Haragan Formation; Devonian; US.

Typical length 1¼ in (3 cm)

| Range: Devonian | Distribution: N. America | Occurrence: ⊛⊛⊛⊛ |

| Group: PHACOPIDA | Subgroup: ENCRINURIDAE | Informal name: Trilobite |

Encrinurus

Especially large tubercles ornament the head shield. The eyes are often stalked, and there are short genal spines. The relatively long tail shield has many more segments in the axis than on the pleural lobes.

HABITAT This trilobite lived in shallow seas in the Silurian.
REMARK *Encrinurus* is known as the "strawberry-headed trilobite" because of its distinctive tubercular head shield.

tubercular head shield

limestone

Encrinurus variolaris Brogniart; Wenlock Limestone; Silurian; UK.

thoracic segment

curved pleural ribs

Typical length 2½ in (6 cm)

| Range: Late Ordovician–Silurian | Distribution: Worldwide | Occurrence: ⊛⊛⊛ |

| Group: LICHIDA | Subgroup: ACIDASPIDAE | Informal name: Trilobite |

Acidaspis

Acidaspis has a wide exoskeleton, with exceptionally long genal spines. Glabellar furrows outline the swollen side lobes. The fairly small eyes are placed at the rear, and small tubercles appear on the head shield, with a fringe of spines around the margin. There are nine narrow thoracic segments, with tubular spines getting longer toward the rear. The rear margin is also fringed with long spines.

HABITAT This trilobite lived in open shelf seas and reefs.
REMARK The extended spines may have been protective.

Typical length
1 in (2.5 cm)

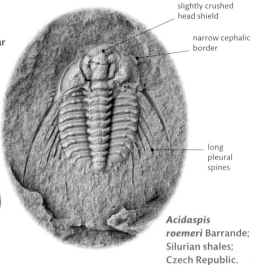

slightly crushed
head shield

narrow cephalic
border

long
pleural
spines

*Acidaspis
roemeri* Barrande;
Silurian shales;
Czech Republic.

| Range: Ordovician–Devonian | Distribution: Worldwide | Occurrence: ⊛⊛⊛ |

| Group: PHACOPIDA | Subgroup: CALYMENIDAE | Informal name: Mudball trilobite |

Flexicalymene

The head shield is convex with a prominent glabella and small eyes. The thoracic segments are downturned and the pygidium is small, similar to that in *Calymene*. These features assist it in enrolling, a likely defensive strategy.

Flexicalymene ouzregui
(Destombes); Late
Ordovician; Morocco.

HABITAT *Flexicalymene* was an inhabitant of warm shelf seas.
REMARK This is a common genus, usually preserved as internal and external molds in limonitic mudstone nodules.

small
eye

large
glabella

thoracic
segments

Typical length
2½ in (6 cm)

**Internal and external
molds**

small
pygidium

| Range: Late Ordovician | Distribution: North America, North Africa | Occurrence: ⊛⊛⊛⊛ |

| Group: PTYCHOPARIIDA | Subgroup: ALOKISTOCARIDAE | Informal name: Trilobite |

Elrathia

Elrathia has a small head compared to its thorax. The glabella is small and flowerpot-shaped with small, centrally placed eyes and short, triangular genal spines. There are 13 narrow segments in the thorax, with a narrow axis and wide pleurae, ending in short, spiny tips. The tail is moderately sized, twice as wide as it is long, with a well-defined axis extending far back.

HABITAT This genus of trilobite swarmed in outer shelf-sea bottoms.
REMARK One of the few species to have been mined commercially, it is often offered for sale.

oval dorsal exoskeleton

furrowed thoracic pleurae

relatively small tail shield

Typical length ¾ in (2 cm)

Elrathia kingii (Meek); Wheeler Shale; Middle Cambrian, US.

| Range: Middle Cambrian | Distribution: N. America | Occurrence: ⊛ ⊛ ⊛ ⊛ ⊛ |

| Group: ASAPHIDA | Subgroup: ASAPHIDAE | Informal name: Trilobite |

Asaphus

The genus *Asaphus* has a smooth, relatively featureless exoskeleton with the exception of its prominent eyes, which are raised on short pedicles or stalks above the head shield. The head and tail shields are semicircular and separated by eight thoracic segments.

HABITAT *Asaphus* lived in a shallow, land-locked inland sea in northeast Europe.
REMARK Eyes on stalks are a feature seen in several unrelated trilobites in this inland sea. This suggests that the bottom water was murky, or possibly it enabled them to watch for predators while buried in the mud.

stalked eye

Asaphus punctatus; (Lessnikova); Middle Ordovician; Russia.

Typical length 1½ in (4 cm)

thoracic segments

virtually smooth pygidium

| Range: Early–Middle Ordovician | Distribution: Northern Europe | Occurrence: ⊛ ⊛ |

Group: CORYNEXOCHIDA	Subgroup: OGYGOPSIDAE	Informal name: Trilobite

Ogygopsis

The exoskeleton of this genus is broadly oval and gently convex, with the axis tapering along its whole length. The near-rectangular glabella extends close to the head margin, with moderately sized eyes in a central position and triangular cheeks. The thorax is made up of eight segments, with deep furrows in the pleurae. The tail shield is exceptionally large and the pleural fields in the tail are strongly furrowed.

HABITAT *Ogygopsis* lived on the bottom of shelf seas.

short genal spines

short, spiny pleural tips

Typical length 3¼ in (8 cm)

Ogygopsis klotzi; Walcott; Burgess Shale, Stephen Formation; Middle Cambrian; Canada.

large tail shield with long axis

axis close to margin

Range: Middle Cambrian	Distribution: N. America, Siberia	Occurrence: ◉

Group: PHACOPIDA	Subgroup: DALMANITIDAE	Informal name: Trilobite

Odontochile

This genus has a semicircular head, small "snout," and long genal spines that extend almost to the tail shield. The thorax has 11 segments that curve backward at their tips. The pygidium is large, subtriangular, deeply furrowed, and terminated by a broad-based spine. The whole surface of the exoskeleton is covered with fine small granules.

HABITAT *Odontochile* inhabited shallow water in warm shelf seas.

large eye

genal spine

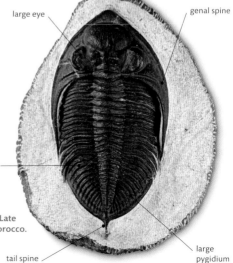

thoracic segments

Typical length 3¼ in (8 cm)

Odontochile hausemann; (Brongniart); Late Devonian; Morocco.

tail spine

large pygidium

Range: Silurian–Devonian	Distribution: Worldwide	Occurrence: ◉ ◉ ◉

CRUSTACEANS

CRUSTACEANS ARE mainly aquatic, carnivorous arthropods whose body and legs are enclosed by a supporting chitinous shell. The body is divisible into head, thorax, and abdomen, but head and thorax may fuse together into a cephalothorax. The head has two pairs of tactile antennae and three pairs of limbs with food-handling abilities. The walking legs are divided in two and may have gills for breathing. The last abdominal appendage may become flattened to form a tail fan with the end spine.

Group: HYMENOSTRACA	Subgroup: HYMENOCARIDIDAE	Informal name: Visored shrimp

Hymenocaris

The near-oval, smooth, bivalved shell of *Hymenocaris* encloses but does not coalesce with the thorax. The thorax has eight segments (somites), each bearing a pair of flattened appendages. The abdomen comprises seven somites and a tail spine (telson) with three pairs of spines–the middle being the longest, the outer very divergent.

HABITAT *Hymenocaris* lived in shallow marine waters.

Typical length 2½ in (6 cm)

Hymenocaris vermicauda Salter; Tremadoc Series; Late Cambrian; UK.

abdominal segments

hingeless, bivalved shell

Range: Cambrian–Ordovician	Distribution: Europe, N. America, Australasia	Occurrence: ◉

Group: THORACICA	Subgroup: BALANIDAE	Informal name: Acorn barnacle

Concavus

This barnacle, shaped like a truncated cone, is made up of six calcareous plates on a solid base. The plates are hollow-walled, but the radii are solid. The operculum, which can be closed by four valves, is the opening through which six pairs of legs (cirri) pass to collect food suspended in water.

HABITAT Barnacles live on rocky shores and floating objects worldwide.

operculum with four plates

symmetrical compartmental plates

Typical length ⅜ in (1 cm)

Concavus concavus (Bronn); Coralline Crag; Pliocene; UK.

base cemented to substrate

Range: Eocene–Recent	Distribution: Worldwide	Occurrence: ◉◉◉◉

Group: DECAPODA	Subgroup: CARPILIIDAE	Informal name: Mud crab

Paleocarpilius

The dorsal carapace of *Paleocarpilius* is egglike in outline, with a slightly extended front, strongly arched transversely, and more steeply rounded longitudinally. The front and side margins are spiny and the claws are robust, with the right one larger. The legs are long and stout. On the ventral surface, the front curves downward and backward to meet the head of the narrow plate in front of the mouth.

HABITAT The genus was largely tropical, living near the shore.

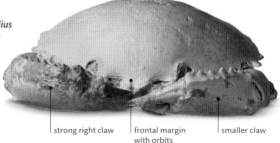

strong right claw | frontal margin with orbits | smaller claw

abdomen with bases of legs

Paleocarpilius aquilinus Collins & Morris; Wadi Thamit Group; Middle Eocene; Libya.

Typical length
2½ in (6 cm)

Range: Eocene–Miocene	Distribution: Europe, Africa	Occurrence: ●●●

Group: THORACICA	Subgroup: STRAMENTIDAE	Informal name: Goose barnacle

Stramentum

Stramentum pulchellum (G. B. Sowerby); Middle Chalk; Late Cretaceous; UK.

Like most cirripedes, the stalked *Stramentum* remained attached as an adult. It is divided into a capitulum, which contains the body, the mouth parts, and thoracic appendages, and a peduncle or stalk, which contains the gonads. The capitulum is protected by 10 calcareous plates. The peduncle is also protected by calcareous plates, but in eight regularly overlapping columns. Each column is surmounted by a plate in the capitulum.

HABITAT These animals frequently lived attached to the empty shells of ammonites, bivalves, or gastropods lying on the sea floor.
REMARK It was once believed that these fossils grew into geese.

Typical length
¾ in (2 cm)

plates enclosing body

plates protecting flexible stalk

point of attachment

Range: Cretaceous	Distribution: Europe, N. Africa, N. America	Occurrence: ●

| Group: PODOCOPIDA | Subgroup: ILYOCYPRIDIDAE | Informal name: Ostracod |

Cyamocypris

The bivalved shell of this creature is thin and nearly oval, with an extended border. The top margin is nearly straight, while the underside is slightly sinuate, converging at the back with the upper margin. The left valve is larger than the right, overlapping on all margins. Each valve bears a marginal notch toward the front and, in typical specimens, a ventral projection known as a beak. The surface of the valves may be pitted, smooth, blistered, or grooved.

HABITAT *Cyamocypris* lived in fresh water.

Typical length
$\frac{1}{50}$ in (0.5 mm)

Cyamocypris valdensis (Fitton); Weald Shales; Early Cretaceous; UK.

Rock containing numerous specimens

Cyamocypris shell

Close-up view

| Range: Early Cretaceous | Distribution: Europe | Occurrence: ◉◉◉◉◉ |

| Group: ISOPODA | Subgroup: Unclassified | Informal name: Sea slater |

Cyclosphaeroma

This creature has a broad body, arched transversely, with a small trilobed head and large eyes near the side margin. Its thorax is composed of eight segments, each with a single-branched appendage. Five abdominal segments are fused as an enlarged triangular tail spine (pleotelson), longitudinally bisected by a median ridge. There are notches on each side of the pleotelson for the insertion of rear abdominal appendages (uropods).

HABITAT An omnivorous scavenger and occasionally a predator, *Cyclosphaeroma* lived at intertidal to moderate depths.

Typical length
$\frac{1}{5}$ in (5 mm)

External mold

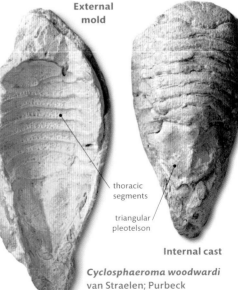

thoracic segments

triangular pleotelson

Internal cast

Cyclosphaeroma woodwardi van Straelen; Purbeck Beds; Late Jurassic; UK.

| Range: Jurassic | Distribution: Europe | Occurrence: ◉◉ |

| Group: DECAPODA | Subgroup: ERYONIDAE | Informal name: Spiny lobster |

Eryon

The carapace is hexagonal and compressed, with sharp side margins. The front is truncated, with spiny margins at the side, and stalked, well-developed compound eyes. The first to fourth thoracic appendages have claws (chelae); the fifth is subchelate. The abdomen is long, flat, and narrow, with a ridge down the middle. It ends in a telson that, with the uropods, forms a tailfan.

HABITAT *Eryon* lived in quiet, clear-water marine lagoons or in shallow water near to the coastline. Today, the family is found only in the ocean depths.

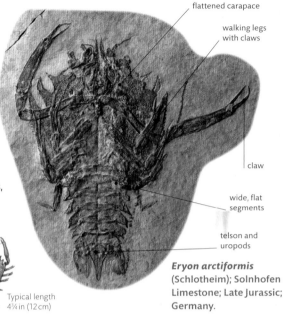

flattened carapace

walking legs with claws

claw

wide, flat segments

telson and uropods

Typical length
4¾ in (12 cm)

Eryon arctiformis
(Schlotheim); Solnhofen
Limestone; Late Jurassic;
Germany.

| Range: Jurassic–Early Cretaceous | Distribution: Europe | Occurrence: ◉ |

| Group: WATERSTONELLIDEA | Subgroup: TEALLIOCARIIDAE | Informal name: Opossum shrimp |

Tealliocaris

The carapace has a single transverse furrow and two prominent longitudinal, keel-shaped ridges (carinae) on each side. The thoracic limbs are double-branched, although with only a single segment in the first pair. The cephalothorax (head and thorax shield) and abdomen are nearly equal in length; the uropods are separated.

HABITAT *Tealliocaris* was non-marine but lived close to the sea.

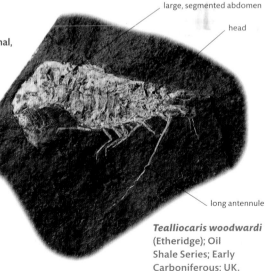

large, segmented abdomen

head

long antennule

Tealliocaris woodwardi
(Etheridge); Oil
Shale Series; Early
Carboniferous; UK.

Typical length
2 in (5 cm)

| Range: Carboniferous | Distribution: Europe | Occurrence: ◉◉◉ |

Group: DECAPODA	Subgroup: PENAEIDAE	Informal name: Shrimp

Acanthochirana

This shrimp has a short, smooth cylindrical carapace and a toothed rostrum. The antennae are longer than the body. The first pair of thoracic appendages are short and articulated. The third pair are the longest. The abdominal segments overlap each other and the tail spine is spindle-shaped.

HABITAT This is a fully marine genus, although some members of the family Penaeidae appear to have lived in brackish and fresh water.

abdominal segment

cylindrical carapace

tailfan

Typical length 1⅜ in (3.5 cm)

Acanthochirana cenomanica Glaessner; Late Cretaceous; Lebanon.

Range: Late Cretaceous	Distribution: Lebanon	Occurrence: ◉◉

Group: DECAPODA	Subgroup: THALASSINIDAE	Informal name: Ghost shrimp

Thalassina

Thalassina has a cylindrical carapace from which a moderately developed rostrum extends. The first and second thoracic appendages are below the claws. The paired appendage arising from the last segment of the body is undivided.

HABITAT *Thalassina* spends most of its time in burrows and is often preserved in hardened burrow infills.

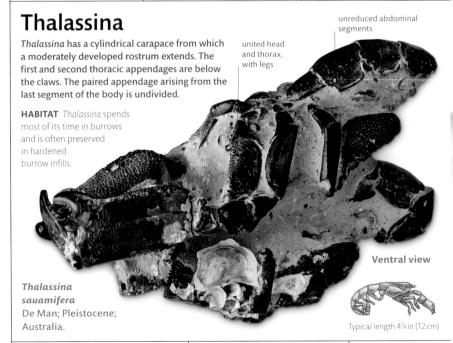

unreduced abdominal segments

united head and thorax, with legs

Ventral view

Thalassina sauamifera De Man; Pleistocene; Australia.

Typical length 4¾ in (12 cm)

Range: Pleistocene–Recent	Distribution: Indo-Pacific	Occurrence: ◉◉

| Group: DECAPODA | Subgroup: PALINURIDAE | Informal name: Spiny lobster |

Linuparus

Closely related to the langouste, *Linuparus* has a carapace compressed from top to bottom without a rostrum, but with three longitudinal ridges. It has spines, also compressed, above the eye sockets and close to the midline. The base of the antenna is fused to the epistome and the side margin. There are no claws on the first four legs, and the fifth leg has a claw only in the female. The uropods and the telson form a broad tailfan.

HABITAT These fossils are found in shallow, marine seas.

Side view

walking leg bases

abdominal segments turned under

very deep cervical groove

abdomen

Linuparus eocenicus Woods; London Clay; Early Eocene; UK.

Typical length 8 in (20 cm)

carapace

segments of abdomen

Dorsal view

| Range: Early Cretaceous–Recent | Distribution: Worldwide | Occurrence: 🔴🔴 |

| Group: DECAPODA | Subgroup: PAGURIDAE | Informal name: Hermit crab |

Pagurus

This hermit crab has a long, weakly calcified carapace and a completely uncalcified abdomen, which consequently coils in the same direction as the host shell (usually a gastropod). The main claws commonly have a right pincer much larger than the left. When not used for walking or scavenging, the claws can be withdrawn to close the aperture of the host shell for protection.

HABITAT It lives in shallow water in temperate seas.
REMARK Complete specimens are rare due to the lack of calcification.

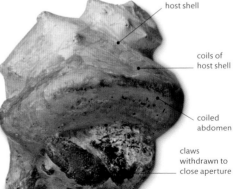

host shell

coils of host shell

coiled abdomen

claws withdrawn to close aperture

Pagurus sp.; Nga Pari Formation; Miocene; New Zealand.

Typical length 1¼ in (3 cm)

| Range: Jurassic–Recent | Distribution: Worldwide | Occurrence: 🔴 |

Group: DECAPODA	Subgroup: PORTUNIDAE	Informal name: Swimming crab

Portunites

Dorsal view

The hexagonal carapace is a little broader than it is long, with four teeth along the front and four or five along the side, of which the most posterior is the longest. The walking legs are as long as the front claws and the fifth leg is unflattened.

HABITAT *Portunites* lived in shallow, warm seas.

Typical length
¾ in (2 cm)

strong, clawed appendage

fifth leg on back

Portunites stintoni
Quayle; London Clay; Early Eocene; UK.

Range: Eocene–Miocene	Distribution: Americas, Europe, Australasia	Occurrence: ◉◉

Group: DECAPODA	Subgroup: GERYONIDAE	Informal name: Mud crab

Archaeogeryon

Archaeogeryon has a near-hexagonal carapace. The eye sockets are large, with straight margins to the side bearing three to five teeth. Toward the rear, margins are long and straight. The massive claws are of different sizes, while the walking legs are strong except for the fifth, which is flattened, suggesting this creature could swim. The abdomen comprises seven segments, with the male abdomen broadly triangular.

HABITAT It is believed that *Archaeogeryon* was a deep-water predator.

Archaeogeryon peruvianus (d'Orbigny); Santa Cruz Beds; Early Miocene; Argentina.

Dorsal view

large carapace

flattened fifth leg

swimming paddle

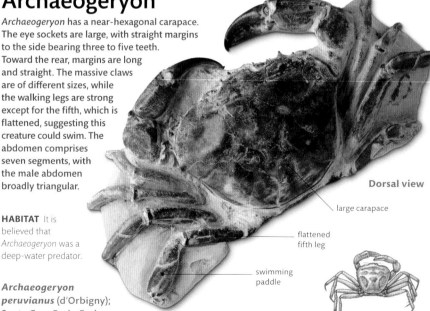

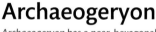

Typical length 6¼ in (16 cm)

Range: Miocene	Distribution: S. America	Occurrence: ◉

CHELICERATES

THE CHELICERATES include horseshoe crabs, spiders, and scorpions. They have bodies divided into a head and thorax shield and an abdomen. Unlike other arthropods, they have no antennae.

Of the six pairs of appendages, the first are claws for feeding (chelicerae), the second are for various functions (pedipalps), and the third to sixth are for walking.

Group: XIPHOSURIDA	Subgroup: EUPROOPIDAE	Informal name: Sea scorpion

Euproops

A shield-shaped carapace covers the front of this sea scorpion, the upper surface being divided into a middle and two side regions by prominent ridges near the simple eyes. The abdomen is composed of seven fused segments bearing marginal spines. The last segment ends in a long, articulating tail spine (telson).

HABITAT *Euproops* is generally believed to be a nonmarine genus.

Euproops rotundatus (Prestwich); Middle Coal Measures; Late Carb.; UK.

fused abdominal segments

Typical length 1½ in (4 cm)

Range: Devonian–Carboniferous	Distribution: N. America, Europe	Occurrence: ◉◉

Group: EURYPTERIDA	Subgroup: EURYPTERIDAE	Informal name: Sea scorpion

Eusarcana

The body of this scorpionlike creature, including its squarish head shield and six pairs of appendages, is encased by a chitinous exoskeleton. Twelve other segments and a final pointed spine articulate with the head shield and with each other. Underneath the head shield are six pairs of appendages: the first with a claw, the second to fifth for walking, and the sixth paddle-shaped.

HABITAT Originally marine, this genus later lived in brackish to fresh water.

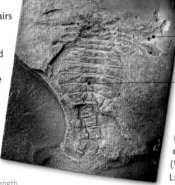

pairs of appendages

abdomen

Eusarcana obesa (Woodward); Ludlow Series; Silurian; UK.

Typical length 4¾ in (12 cm)

Range: Silurian	Distribution: Europe, N. America, Asia	Occurrence: ◉

Group: XIPHOSAURIDA	Subgroup: LIMULIDAE	Informal name: Horseshoe crab

Mesolimulus

This is the precursor of the living horseshoe crab, so called because the carapace is horseshoe shaped. The upper surface of the carapace is smooth, except for a central ridge and two longitudinal ridges. The eyes are small and widely spaced, lying just outside the longitudinal ridges. The abdomen is unsegmented, but six short spines are ranged around the margin. The abdomen articulates with the carapace and posteriorly with the long, sharp tail spine (telson).

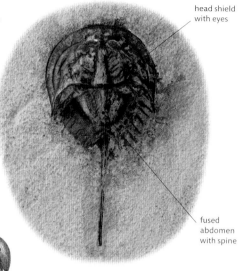

head shield with eyes

fused abdomen with spine

HABITAT Modern horseshoe crabs are common along the eastern seaboard of the US, in the Indian Ocean, and in southeast Asia. They are tolerant of changes in salinity and migrate to lay eggs in shallow, intertidal mud flats. They are relatively omnivorous, living on seaweeds, dead fish, and small crustaceans.

Typical length
4¾ in (12 cm)

Mesolimulus walchii (Desmarest); Solnhofen Limestone; Late Jurassic; Germany.

Range: Jurassic–Cretaceous	Distribution: Europe, M. East	Occurrence: ◉◉

Group: SCORPIONES	Subgroup: PARAISOBUTHIDAE	Informal name: Water scorpion

Paraisobuthus

This aquatic scorpion has abdominal plates divided into two lobes. The first two pairs of coxae (the leg segments nearest to the body) are greatly enlarged into maxillary lobes for feeding. The carapace is squarish, with two well-developed cheeks separated by a deep groove. The stinger is very similar in shape to those of Recent terrestrial scorpions.

powerful claws

walking legs

body

long tail

stinger

HABITAT These animals were largely aquatic, being able to leave water for only a short time. They are presumed to have been carnivorous, as are modern scorpions, but the stinger may have been used for defense, as well as attack.

Paraisobuthus prantli Kjellesvig-Waering; Radnice Group; Late Carboniferous; Czech Republic.

Typical length
2¾ in (7 cm)

Range: Late Carboniferous	Distribution: N. America, Europe	Occurrence: ◉

Group: AMBLYPYGI	Subgroup: PARACHARODONTIDAE	Informal name: "Spider"

Graeophonus

This genus, which is close to true spiders, has an undivided carapace, with a distinct frontal projection. A single pair of round eyes sits on sharply defined, knoblike projections. To the front is a pair of slender, tactile feeding claws (chelicerae), with a powerful first walking leg adapted to seize prey. The abdomen is composed of 12 segments, and the legs are segmented like those of modern spiders.

HABITAT *Graeophonus* lived in shallow-water swamps, deltas, or lagoons.

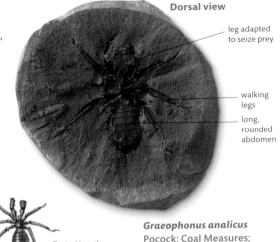

Dorsal view

leg adapted to seize prey

walking legs

long, rounded abdomen

Typical length
⅝ in (1.5 cm)

Graeophonus analicus Pocock; Coal Measures; Late Carboniferous; UK.

Range: Late Carboniferous	Distribution: Europe, N. America	Occurrence: ◉◉

Group: ARANEAE	Subgroup: PISAURIDAE	Informal name: Spider

Dolomedes

The cephalothorax of this true spider is slightly longer than it is broad and high and convex near the eyes. Set in two rows, the eyes in the posterior row are larger. The legs are strongly developed and of almost equal length except for the third pair, which is shorter.

HABITAT *Dolomedes* lives near fresh water, running on the surface to capture its insect prey.

Specimen in kauri resin

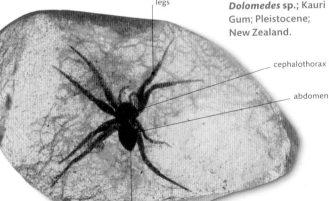

legs

Dolomedes sp.; Kauri Gum; Pleistocene; New Zealand.

cephalothorax

abdomen

spinnerets for filament secretion

Typical length
1 in (2.5 cm)

Range: Pleistocene–Recent	Distribution: Worldwide	Occurrence: ◉

INSECTS

INSECTS ARE TERRESTRIAL and freshwater arthropods, with greater diversity of species than any other animal. Their bodies are divided into head, thorax, and abdomen, with three pairs of walking legs attached to the thorax. The earliest insects, found in the Early Devonian, are wingless. However, the majority are winged, and, by the Late Carboniferous, this group became the first animals to evolve powered flight. Most insects undergo a metamorphosis, with a pupa as a resting stage.

Group: ODONATA	Subgroup: PETALURIDAE	Informal name: Dragonfly

Petalura

This dragonfly has four broad, richly veined wings. The body is long and slender.

HABITAT The young (nymphs) live in fresh water, feeding on other aquatic animals, while the adults hunted flying insects near fresh water.

REMARK The reconstructed species shown is a true dragonfly, but its relationship with the Recent *Petalura* is open to question.

lithographic limestone

forewing

hindwing

abdomen

Petalura sp.; Solnhofen Limestone; Late Jurassic; Germany.

Typical length 3 in (7.5 cm)

Range: Jurassic–Recent	Distribution: Europe, Australasia	Occurrence: ◉◉

Group: BLATTOPTERA	Subgroup: ARCHIMYLACHRIDAE	Informal name: Cockroach

Archimylacris

Archimylacris was an early terrestrial cockroach with folded wings, a large head shield (pronotum), and strengthened forewings (tegmina).

HABITAT *Archimylacris* lived in warm, moist forests and was probably an omnivorous scavenger.

REMARK In this specimen, the wings are well preserved. This quality of preservation enables paleontologists to determine that the specimen is an adult; immature stages do not have full wings.

ironstone nodule

thorax

forewing

hindwing

Typical length 1¼ in (3 cm)

Archimylacris eggintoni (Bolton); Coal Measures; Late Carboniferous; UK.

Range: Late Carboniferous	Distribution: Europe, N. America	Occurrence: ◉◉◉

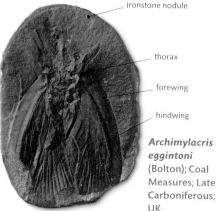

| Group: PLECOPTERA | Subgroup: MESOLEUCTRIDAE | Informal name: Stonefly |

Mesoleuctra

Mesoleuctra was a slender, aquatic nymph with a terrestrial adult, with folded wings. The nymph had two tails (cerci) and four wing pads.

HABITAT Adults lived among stones and foliage near the streams, rivers, and lakes in which the nymphs lived. The stonefly's diet was probably small plants, such as lichens and small algae. Today, stonefly nymphs are an important source of food for freshwater fish.

REMARK This specimen was probably buried very quickly in lake sediments, thus preserving the fragile legs. Fossil aquatic insects are relatively rare. More commonly, insect-bearing rocks contain terrestrial species blown into lake sediments.

Typical length
1 in (2.5 cm)

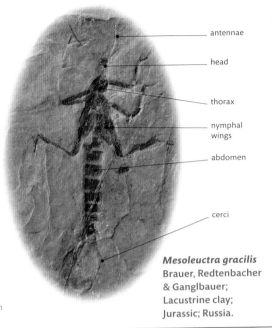

antennae
head
thorax
nymphal wings
abdomen
cerci

Mesoleuctra gracilis
Brauer, Redtenbacher & Ganglbauer;
Lacustrine clay;
Jurassic; Russia.

| Range: Early Jurassic–Recent | Distribution: Asia | Occurrence: ●●● |

| Group: COLEOPTERA | Subgroup: HYDROPHILIDAE | Informal name: Water beetle |

Hydrophilus

The forewings of *Hydrophilus* are modified into hindwing cases (elytra) in the adult. These, along with the abdominal hairs, trap air and act as air reservoirs underwater.

HABITAT Both young and adult live in fresh water, feeding on plant and animal matter. The larvae are carnivorous.

REMARK This specimen was mired in asphalt, which had naturally seeped into a shallow lake in sandy wetland. Such "tar" pits preserved a wealth of Pleistocene animals in California, including imperial mammoths and saber-toothed cats.

Typical length
1¼ in (3 cm)

head
thorax
elytra
tar sand

Hydrophilus
sp.; Tar sands;
Pleistocene; US.

| Range: Pliocene–Recent | Distribution: N. America, Europe | Occurrence: ●●●●● |

| Group: COLEOPTERA | Subgroup: OMMATIDAE | Informal name: Beetle |

Blapsium

This beetle has folded wings with roughly sculptured elytra (compare *Hydrophilus*, p.77). The basal leg segments (coxae) of the hindlegs evidently do not divide the first ventral plate (sternite) of the abdomen. The coxae, which were attached to the thorax by muscles, have dropped out, leaving cavities.

HABITAT *Blapsium* was an omnivorous scavenger that lived close to the sea. It is found fossilized alongside marine shells.

REMARK The Stonesfield Slate is famous for its fossils, which are a mixture of marine dwellers, land animals, and plants washed out to sea.

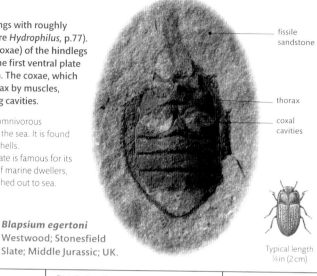

fissile sandstone

thorax

coxal cavities

Blapsium egertoni
Westwood; Stonesfield Slate; Middle Jurassic; UK.

Typical length ¼ in (2 cm)

| Range: Jurassic | Distribution: Europe | Occurrence: ◉ |

| Group: DIPTERA | Subgroup: BIBIONIDAE | Informal name: March fly |

Bibio

This insect has only one pair of wings (forewings), folded and developed internally, with the hindwings reduced to balance organs (halteres), characteristic of true flies. The forewings have reduced vein patterns and pigmentation.

HABITAT March flies typically inhabit grassland, often appearing in spring (hence their name) and visiting flowers.

limestone

Bibio maculatus
Heer; Miocene; Croatia.

forewing

abdomen

leaf impression

Typical length ⅜ in (1 cm)

| Range: Pliocene–Recent | Distribution: Europe | Occurrence: ◉◉ |

BRACHIOPODS

ALTHOUGH LIVING SPECIES are now rare, brachiopods are common in marine fossiliferous rocks. Over 3,000 genera have been described, from the Cambrian to Recent times, and they form the most abundant fauna in many Paleozoic rocks. The shell comprises two valves: the pedicle valve on the ventral side and the smaller brachial valve on the dorsal side. A stalk (pedicle) usually emerges from the rear of the pedicle valve, attaching the brachiopod to the sea floor. The two main groups—the hinged Articulata and the hingeless Inarticulata—are further divided into 12 orders.

Group: LINGULIDA	Subgroup: LINGULIDAE	Informal name: Brachiopod

Lingula

This relatively small, inarticulate brachiopod has a thin shell, with both valves gently convex and tongue-shaped in outline. The beak of the pedicle valve has a triangular groove for the supporting pedicle. Internally, there are no teeth or sockets. The ornament consists of fine growth lines only.

HABITAT Like all Recent lingulids, the fossil *Lingula* probably lived in vertical burrows in intertidal areas.
REMARK This "living fossil" has changed little in 400 million years.

apex (umbo)

calcium phosphate shell

fine growth lines

Typical length
⅝ in (1.5 cm)

Lingula credneri
Greinitz; Marl Slate Formation; Permian; UK.

Range: Ordovician–Recent	Distribution: Worldwide	Occurrence: ●●●●●

Group: ORTHIDA	Subgroup: PLATYSTROPHIDAE	Informal name: Brachiopod

Platystrophia

The shell of *Platystrophia* is near-rectangular in outline, with strongly convex valves. The shell reaches its maximum width along the hinge line, and the hinge often extends to form sharp "ears" (auricles). A conspicuous depression (sulcus) is found on the pedicle valve, with a corresponding fold on the brachial valve. The line of junction between the shells is strongly folded at the front edge. Numerous sharp-crested ribs diverge toward the shell margins. *Platystrophia* was a medium-sized brachiopod.

HABITAT *Platystrophia* was attached by a short pedicle to lime-mud and sandy substrates.

maximum width along hinge line

Platystrophia biforata (Schlotheim); Hudson River Group; Ordovician; US.

Typical length
1½ in (4 cm)

deep fold

sharp-crested ribs

Range: Ordovician	Distribution: Worldwide	Occurrence: ●●●●

Group: ORTHIDA	Subgroup: DICOELOSIIDAE	Informal name: Brachiopod

Dicoelosia

The shell of this brachiopod has a characteristic bilobed, near-triangular outline, with a strongly convex pedicle valve and less convex brachial valve. Both valves have a pronounced sulcus, with a strongly incurved beak on the pedicle valve. The hinge line is of variable width but is generally short. The interarea (between the valves) is medium-sized, with an open groove for passage of the pedicle. Numerous fine ribs diverge near the front of the shell.

HABITAT *Dicoelosia* was attached to bryozoans and shelly fragments in shallow to mid-depths.

twin-lobed pedicle valve

Dicoelosia bilobata (Linnaeus); Wenlock Limestone; Silurian; UK.

Typical length
⅝ in (1.5 cm)

deep sulcus

ornament of fine ribs

Range: Ordovician–Devonian	Distribution: Worldwide	Occurrence: ⓐⓐⓐⓐⓐ

Group: STROPHOMENIDA	Subgroup: STROPHOMENIDAE	Informal name: Brachiopod

Leptaena

This articulate brachiopod had a shell with a semicircular outline, prominent auricles, and a straight hinge line. The pedicle valve is slightly convex and the brachial valve usually flat except at the front margins, where both valves bend almost at right angles. The surface ornament consists of numerous fine, clustered ribs and strong, concentric wrinkles (rugae).

HABITAT *Leptaena* lay on the sea floor with its pedicle valve downward. The upward-projecting front margin was kept clear of the sediment surface, allowing the brachial valve to be almost completely buried.

REMARK It is thought that the concentric rugae stabilized the shell in soft substrates. *Leptaena* is usually found with both valves separated in fine-grained limestone shales.

hinge line

Leptaena sp.; Longhope Formation; Silurian; US.

Typical length
1½ in (4 cm)

sharp bends at front margin

clustered ribs

strong rugae

Range: Ordovician	Distribution: Worldwide	Occurrence: ⓐⓐⓐⓐⓐ

Group: PRODUCTIDA	Subgroup: PRODUCTIDAE	Informal name: Brachiopod

Productus

The shell has a flat to concave brachial valve that is nearly circular in outline. The pedicle valve is thick and strongly convex, with a short hinge line and long trail (the extension of the valve beyond the mantle cavity). The ornament consists of numerous ribs and irregular wrinkles. Scattered spines can be found on the pedicle valve but are usually preserved only as spine bases. No spines are found on the brachial valve.

HABITAT *Productus* lived in soft, muddy sediments, its thicker pedicle valve partially buried.
REMARK The spines stabilized the shell, preventing it from sinking too far into the sediment. The long trail probably projected vertically upward, keeping the valve margins clear of the substrate.

Productus productus (Martin); Carboniferous Limestone; Early Carboniferous; UK.

ribs

rugae

pedicle valve

trail

Typical length
1¾ in (4.5 cm)

Range: Carboniferous	Distribution: Worldwide	Occurrence: ◉◉◉◉◉

Group: STROPHOMENIDA	Subgroup: ECHINOCONCHIDAE	Informal name: Lump shell

Parajuresania

The shell of this medium-sized brachiopod has a flat to slightly concave brachial valve and a pedicle valve that is strongly convex, with a short trail and shallow middle depressions. Both valves are strongly ornamented by concentrically arranged, overlapping rows of spines, with concentric ridges more prominent near the front.

HABITAT *Parajuresania* lay in soft sediment.

Parajuresania symmetrica (McChesney); Finis Formation; Late Carboniferous; US.

nearly erect spines

Typical length
1¼ in (3 cm)

prostrate spines

concentric wrinkles

Range: Late Carboniferous	Distribution: Worldwide	Occurrence: ◉◉◉

Group: PENTAMERIDA	Subgroup: PENTAMERIDAE	Informal name: Brachiopod

Pentamerus

In this relatively large brachiopod, both pedicle and brachial valves are convex and often tongue-shaped in outline. The pedicle valve umbo (apex area) is stout and curved strongly backward, with a prominent middle partition (septum). Two prominent, near-parallel septa are found on the brachial valve. The shell exterior is smooth, with fine growth lines.

HABITAT *Pentamerus* is very common in shallow-water limestones, often forming shell banks. The pedicle was not functional in adults, so they lived umbones downward on silty and muddy substrates.

pedicle valve

near-parallel septa

line of junction

smooth shell

Typical length
1¼ in (4.5 cm)

Pentamerus oblongus
(J. de C. Sowerby); Jupiter Formation; Silurian; Canada.

Range: Silurian	Distribution: Worldwide	Occurrence: ◉◉◉◉◉

Group: RHYNCHONELLIDA	Subgroup: PUGNACIDAE	Informal name: Brachiopod

Pugnax

The shell of this small to large brachiopod has a distinctive four-sided outline. The brachial valve is globelike, with a pronounced fold at the front. In juvenile forms, the pedicle valve is convex, but in large individuals, it becomes concave toward the shell margins, forming a deep and nearly rounded sulcus. Apart from a few faint ribs, there is almost no ornament on either the brachial or pedicle valve.

HABITAT *Pugnax* was tethered by a functional pedicle to shelly fragments or other hard surfaces. It is commonly found in large clusters, especially in Carboniferous mud mounds and shallow-water reef mounds.

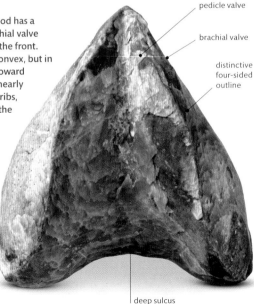

pedicle valve

brachial valve

distinctive four-sided outline

deep sulcus

Typical length
1¼ in (4.5 cm)

Pugnax acuminatus
(J. Sowerby); Carboniferous Limestone; Early Carboniferous; UK.

Range: Devonian–Late Carboniferous	Distribution: Europe	Occurrence: ◉◉◉◉

Group: RHYNCHONELLIDA	Subgroup: PUGNACIDAE	Informal name: Brachiopod

Pleuropugnoides

A relatively small but distinctive brachiopod, *Pleuropugnoides* had a shell that is nearly triangular in outline. The pedicle valve is convex at the back but concave toward the front, developing a large sulcus. The brachial valve is globular, with a broad front fold. The line of junction between the valves (commissure) undulates strongly at the fold crest. Sharp, diverging ribs ornament the shell.

HABITAT *Pleuropugnoides* lived on Carboniferous reefs.

Pleuropugnoides pleurodon (J. Phillips); Carboniferous Limestone; Early Carboniferous; UK.

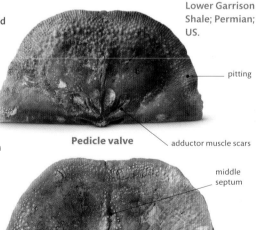

pedicle valve

sharp ribs

fold in brachial valve

undulating commissure

Typical length ⅝ in (1.5 cm)

Range: Early–Late Carboniferous	Distribution: Europe	Occurrence: ●●●●

Group: PRODUCTIDA	Subgroup: CHONETIDAE	Informal name: Brachiopod

Chonetes

The shell of this relatively small chonetid has a semicircular outline and a long hinge, which extends to form auricles. The pedicle valve is convex; the brachial valve is flat to slightly concave. Internally, the teeth and sockets are weakly developed. A thin middle septum occurs on the pedicle valve, along with distinct, two-lobed muscle scars. The brachial valve also has a middle septum and back and front muscle scars. Both valves are pitted, with ribs and short spines ranged along the rear edge of the narrow interarea.

HABITAT *Chonetes* lived on soft, lime-mud substrates.

Chonetes sp.; Lower Garrison Shale; Permian; US.

pitting

Pedicle valve

adductor muscle scars

middle septum

auricle

Typical length ¾ in (2 cm)

Brachial valve

short spines

Range: Carboniferous–Permian	Distribution: Worldwide	Occurrence: ●●●●

Group: RHYNCHONELLIDA	Subgroup: CYCLOTHYRIDIDAE	Informal name: Brachiopod

Cyclothyris

The shell of this relatively large and conspicuous rhynchonellid is characterized by a wide, near-triangular outline. The beak is erect and contains a large hole (foramen) for the pedicle. Both valves are convex, with a pronounced sulcus on the pedicle valve and a fold on the brachial valve. The front junction line zigzags, and numerous small, sharp-crested ribs ornament both valves.

HABITAT *Cyclothyris* is commonly found in sediments that have been deposited in moderate- to high-energy environments.

Cyclothyris difformis (Valenciennes); Lower Chalk; Late Cretaceous; UK.

foramen

Typical length 1¼ in (3 cm)

brachial valve

sharp-crested ribs

fold

Range: Cretaceous	Distribution: Europe, N. America	Occurrence: ●●●

Group: ATRYPIDA	Subgroup: ATRYPIDAE	Informal name: Brachiopod

Atrypa

The brachial valve of this genus is more convex than the pedicle valve and has a broad, shallow fold. The pedicle valve has a small depression and a slight, curved-in beak. The front junction line is often slightly deflected toward the brachial valve. The ornament consists of numerous fine, rounded ribs and prominent, concentric growth lines commonly forming frills on both valves.

HABITAT *Atrypa* lived in shallow water on soft substrates.

pedicle valve

fine ribs

Typical length ½ in (1.2 cm)

Atrypa sp.; Eke Beds; Silurian; Sweden.

deflected junction line

growth-line frills

Range: Silurian–Late Devonian	Distribution: Worldwide	Occurrence: ●●●●

Group: ATHYRIDIDA	Subgroup: ATHYRIDIDAE	Informal name: Brachiopod

Actinoconchus

The valves of this distinctive brachiopod are nearly equally convex, and the shell outline is near-circular. Specimens commonly lack a well-developed fold or depression. Broad, thin-layered expansions are developed at growth lines. These are traversed by a series of fine ribs that diverge toward the front margins of the shell.

HABITAT *Actinoconchus* was attached by a short, functional pedicle to hard substrates. The shape of the brachial valve commonly conforms to that of the surface of the substrate.

REMARK The thin-layered expansions on *Actinoconchus* served as camouflage and prevented the shell from sinking into muddy sediments.

brachial valve

ribs

growth-line expansions

line of junction

Typical length
1¾ in (4.5 cm)

Actinoconchus paradoxus (McCoy); Carboniferous Limestone; Early Carboniferous; Ireland.

Range: Early Carboniferous	Distribution: Europe	Occurrence: ◉◉◉

Group: ATHYRIDIDA	Subgroup: MERISTELLIDAE	Informal name: Brachiopod

Meristina

The shells of these relatively large brachiopods are triangular to round in shape and are often longer than they are wide. The brachial valve has a broad, shallow fold, with a corresponding sulcus on the pedicle valve. The apex areas on both valves are strongly curved in, concealing the pedicle opening in large forms. Both valves are unornamented.

HABITAT *Meristina* is found in small clusters in fine-grained, muddy limestones. It probably lived in shallow water.

fold in brachial valve

umbo

Typical length
1¾ in (4.5 cm)

sulcus in pedicle valve

Meristina obtusa (J. de C. Sowerby); Wenlock Limestone; Silurian; UK.

Range: Silurian–Devonian	Distribution: Worldwide	Occurrence: ◉◉◉◉

| Group: SPIRIFERIDA | Subgroup: MUCOSPIRIFERIDAE | Informal name: Butterfly shell |

Mucrospirifer

The shell of this relatively large spiriferid reaches its maximum width along the hinge line, which is extended outward to form sharp points. The brachial valve has a pronounced fold, with a corresponding depression on the pedicle valve. The apex area of the pedicle valve contains a small hole for the pedicle.

HABITAT *Mucrospirifer* lived in soft, muddy substrates.

Mucrospirifer mucronata (Conrad); Hamilton Group; Devonian; Canada.

umbo of pedicle valve

ribs

alate hinge line separating inhalant and exhalant currents

regularly overlapping growth layers

Typical length 1 in (2.5 cm)

| Range: Devonian | Distribution: Worldwide | Occurrence: ●●● |

| Group: ORTHIDA | Subgroup: SCHIZOPHORIIDAE | Informal name: Brachiopod |

Aulacophoria

In outline, the shell of this medium-sized orthid varies from nearly circular to nearly square. The brachial valve is more convex than the pedicle valve and has a broad front fold, corresponding to the depression on the pedicle valve. The pedicle itself is prominent, and the interarea relatively narrow. Both valves are ornamented by many fine ribs.

HABITAT *Aulacophoria* lived in lime-mud sediments.

pedicle valve

small interarea

fold in brachial valve

Typical length 1¼ in (3 cm)

Aulacophoria keyserlingiana (de Koninck); Yoredales; Early Carboniferous; UK.

| Range: Early Carboniferous | Distribution: Europe | Occurrence: ●●● |

Group: TEREBRATULIDA	Subgroup: POSTEPITHYRIDIDAE	Informal name: Lamp shell

Epithyris

The shell of this medium-sized to large terebratulid has a near-circular outline. Its convex pedicle valve forms an incurved erect beak, which contains a large pedicle opening (foramen) with a well-developed collar. The front junction line undulates, with two folds on the brachial valve and a corresponding depression (sulcus) on the pedicle valve. The folds can be sharp-crested or rounded. There is little surface ornament— only fine growth lines which can become incised and attenuated near the shell's front edges.

HABITAT It is likely that *Epithyris* lived in relatively quiet, lagoonal waters.
REMARK These brachiopods formed communal "nests," sometimes extending for several feet.

Epithyris maxillata (J. de C. Sowerby); Bradford Clay; Middle Jurassic; UK.

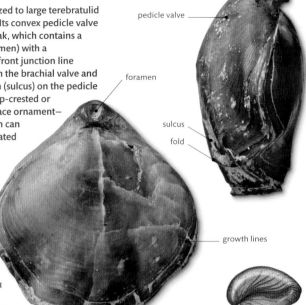

erect beak
pedicle valve
foramen
sulcus
fold
growth lines
brachial valve

Typical length 1¾ in (4.5 cm)

Range: Jurassic	Distribution: Europe	Occurrence: ◉◉◉◉◉

Group: TEREBRATULIDA	Subgroup: DICTYOTHYRIDAE	Informal name: Lamp shell

Dictyothyris

The shell of this distinctive, small- to medium-sized terebratulid has a five-sided outline. The convex pedicle and brachial valves have an erect beak and large pedicle opening. Two middle folds on the brachial valve and corresponding structures on the pedicle valve produce a W-shaped junction at the front. The valves are ornamented with five longitudinal and transverse ribs, with spinules.

HABITAT *Dictyothyris* lived attached to shelly fragments in soft, muddy sediments.

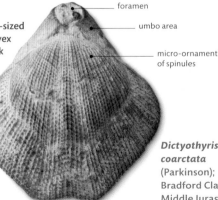

foramen
umbo area
micro-ornament of spinules

Dictyothyris coarctata (Parkinson); Bradford Clay; Middle Jurassic; UK.

Typical length ¾ in (2 cm)

Range: Jurassic	Distribution: Europe	Occurrence: ◉◉◉◉◉

Group: STROPHOMENIDA	Subgroup: STROPHOMENIDAE	Informal name: Brachiopod

Strophomena

The shell of this medium-sized brachiopod has a convex brachial valve and a concave pedicle valve, with the maximum width along the hinge line. Internally, the brachial valve has well-developed sockets and a prominent cardinal process (the projection from the hinge line to which the diductor muscles are attached). The hole for the pedicle is small and partially covered by a plate. Both valves are ornamented by fine ribs, which diverge toward the front of the shell.

HABITAT The small foramen indicates that *Strophomena* was probably free-lying, resting with its brachial valve downward on a variety of soft substrates.

Strophomena grandis (J. de C. Sowerby); Cheney Longville Formation; Ordovician; UK.

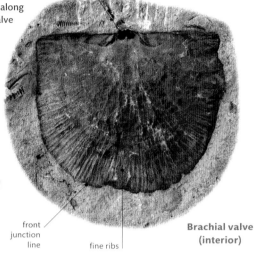

well-developed sockets

Typical length
1¼ in (3 cm)

front junction line

fine ribs

Brachial valve (interior)

Range: Ordovician	Distribution: Worldwide	Occurrence: ⬤⬤⬤

Group: SPIRIFERINIDA	Subgroup: CYRTINIDAE	Informal name: Brachiopod

Cyrtina

The shell of this small- to medium-sized spiriferid has a characteristic caplike shape and maximum width along the hinge line. On the pedicle valve, there is a large, triangular interarea, with a large hole for the pedicle near the apex and a distinct sulcus that widens toward the front. The brachial valve is weakly convex, with an unornamented fold and depression. Several distinct ribs on both valves produce an undulating junction line, and growth lines become prominent near the front margins.

HABITAT *Cyrtina* lived in soft sediments.

sulcus on pedicle valve

distinctive caplike shape

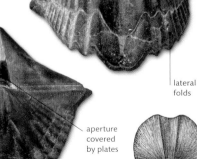

lateral folds

prominent growth lines

aperture covered by plates

Cyrtina hamiltonensis (Hall); Hamilton Group; Devonian; Canada.

Typical length
1⅞ in (5 cm)

Range: Devonian–Permian	Distribution: Worldwide	Occurrence: ⬤⬤⬤⬤

Group: ORTHOTETIDA	Subgroup: DERBYIIDAE	Informal name: Brachiopod

Derbyia

The shell of this medium-sized strophomenid is semicircular in outline, with convex valves. A shallow depression may appear in the pedicle valve, the depth of which may vary across the shell. The short, stubby apex area is sometimes offset. Numerous fine, radiating ribs and prominent growth lines ornament the shell.

HABITAT It is likely that *Derbyia* was free-living.

short umbo

Derbyia grandis (Waagen); Middle Productus Limestone; Early Permian; India.

ornament of fine ribs

growth lines

Pedicle valve

Typical length 1⅜ in (3.5 cm)

Range: Early Carboniferous–Late Permian	Distribution: Worldwide	Occurrence: ⊛⊛⊛⊛

Group: SPIRIFERINIDA	Subgroup: SPIRIFERINIDAE	Informal name: Brachiopod

Spiriferina

The shell of *Spiriferina* is triangular to near-pentagonal in outline, with convex valves. The umbones are strongly curved in, with a prominent pedicle and well-developed interarea. A deep fold is found on the brachial valve and a deep sulcus on the pedicle valve. The line of junction undulates strongly, and both valves are ornamented by large, rounded ribs. Well-preserved specimens often display a micro-ornament of fine spines.

HABITAT *Spiriferina* lived with its umbones downward on soft, muddy sediments.

umbo

Pedicle valve

sulcus

interarea

growth lines

undulating junction line

ribs

Typical length ⅝ in (1.5 cm)

Spiriferina walcotti (J. Sowerby); Lower Lias; Early Jurassic; UK.

fold

Brachial valve

Range: Triassic–Early Jurassic	Distribution: Worldwide	Occurrence: ⊛⊛⊛⊛⊛

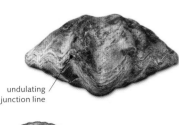

Group: RHYNCHONELLIDA	Subgroup: TETRARHYNCHIIDAE	Informal name: Brachiopod

Goniorhynchia

The medium-sized shell of this brachiopod has a near-triangular outline. The convex pedicle valve has a prominent beak and a well-developed sulcus, while the fold in the brachial valve is most prominent near the front of the shell. A strongly undulating junction line is highlighted by attenuated growth lines. Both valves are ornamented by numerous sharp-crested ribs.

HABITAT This was a solitary brachiopod that rested with its umbones downward on firm substrates. The pedicle acted as a tether.

REMARK A typical thick-shelled rhynchonellid, *Goniorhynchia* can be very abundant in certain areas.

prominent beak on pedicle valve

brachial valve

fold

ornament of sharp-crested ribs

Typical length
¾ in (2 cm)

Goniorhynchia boueti
(Davidson); Boueti Bed; Forest Marble; Middle Jurassic; UK.

Range: Middle Jurassic	Distribution: Europe	Occurrence: ⦾⦾⦾⦾⦾

Group: TEREBRATULIDA	Subgroup: ZEILLERIIDAE	Informal name: Lamp shell

Digonella

The shell of this relatively small- to medium-sized terebratulid is long and oval to sack-shaped in outline. Both valves are convex, and the shell reaches its maximum width near the front. The apex area is flattened toward the back, and the pedicle is curved and short, with a large opening (foramen) for the stalk to pass through. The front line of junction is straight, and there is no ornament other than the very fine growth lines.

HABITAT *Digonella* lived in soft sediment, predominantly lime mud, where it was attached to shell fragments by a short pedicle. It is abundant in shallow-water, fine-grained limestones.

foramen

pedicle valve

brachial valve

straight line of junction

Typical length
1 in (2.5 cm)

Digonella digona
(J. Sowerby); Bradford Clay; Middle Jurassic; UK.

Range: Middle Jurassic	Distribution: Worldwide	Occurrence: ⦾⦾⦾⦾

Group: TEREBRATULIDA	Subgroup: TEREBRATALIIDAE	Informal name: Lamp shell

Terebrirostra

This genus is distinguished by a long, curving umbo, which extends behind the rear margin of the shell. The brachial valve is oval to near-triangular in outline. Along the front line of junction between the valves are many undulations and depressions. Both the brachial and pedicle valves are ornamented by radial ribs and fine, concentric growth lines.

HABITAT *Terebrirostra* was probably a solitary brachiopod attached to hard substrates by a short, functional pedicle. The conspicuous curved umbo on the pedicle valve may have kept the shell clear of the substrate, preventing sediment from entering the mantle cavity. This genus is fairly common in shallow-water sediments, particularly glauconite chalks.

long, curving umbo

interarea

ornament of ribs and fine growth lines

brachial valve

Typical length 1½ in (4 cm)

Terebrirostra bargensa (d'Orbigny); Lower Chalk; Late Cretaceous; France.

Range: Late Cretaceous	Distribution: Europe	Occurrence: ◉◉

Group: TEREBRATULIDA	Subgroup: KINGENIDA	Informal name: Lamp shell

Kingena

This medium-sized terebratulid had a shell with convex valves. The shell is characterized by a rounded to near-pentagonal outline, with a broad, shallow fold in the brachial valve and a faint depression on the pedicle valve. The pedicle itself is short and curves slightly backward. The brachial valve commonly exhibits a prominent middle partition (septum). Fine growth lines and minute granules ornament the shell.

HABITAT *Kingena* was attached to hard substrates by a short, thick, fleshy pedicle. This is a common brachiopod on shallow-water limestones.

hole for passage of pedicle

middle septum

brachial valve

fine growth lines

Typical length 1¼ in (3 cm)

Kingena lemanensis (Pictet & Le Roux); Shenley Limestone; Early Cretaceous; UK.

Range: Cretaceous	Distribution: Worldwide	Occurrence: ◉◉◉◉

Group: CRANIIDA	Subgroup: CRANIIDAE	Informal name: Brachiopod

Danocrania

The shell of this small brachiopod is almost circular in outline, with a straight rear margin. The shape of the pedicle valve conforms to the underlying substrate. Both valves exhibit a well-developed pair of circular muscle scars, accompanied by smaller muscle scars, and both thicken at their front margins. Apart from surface blisters (pustules), the ornament consists of concentric growth lines, and may have provided some form of camouflage.

Danocrania tuberculata (Nilsson); Danian chalk; Paleocene; Denmark.

HABITAT Most genera in the family are cemented to hard substrates by their pedicle valves.

REMARK The family appeared in the Early Mesozoic and is still alive in today's oceans. They are unique in lacking a pedicle.

Typical length
⅜ in (1 cm)

pustulose ornament

straight rear margin

circular muscle scars

Range: Early Paleocene	Distribution: Worldwide	Occurrence: ◉◉◉◉◉

Group: LINGULIDA	Subgroup: DISCINIDAE	Informal name: Brachiopod

Discinisca

The genus is characterized by a gently conical brachial valve and a flat to slightly convex pedicle valve. The often shiny shell is near-circular to spatulate in outline, with a thin, slitlike pedicle opening that is closed in adult forms and open at the rear in juveniles. The ornament consists of fine, concentric growth lines, prominent near the front margin of the shell.

conical brachial valve

position of apical opening

HABITAT This genus is commonly found in its life position attached to a variety of hard substrates, particularly other shells.

REMARK Some genera from the family Discinidae are "living fossils."

calcium phosphate shell

Typical length
⅓ in (8 mm)

accentuated growth lines

Discinisca lugubris (Conrad); Choptank Formation; Miocene; US.

Range: Permian–Recent	Distribution: Worldwide	Occurrence: ◉◉◉◉

| Group: TEREBRATULIDA | Subgroup: CANCELLOTHYRIDIDAE | Informal name: Lamp shell |

Cancellothyris

This medium-sized to relatively large brachiopod is characterized by an egg-shaped shell, with convex valves and a short but massive apex area that is truncated by a large pedicle opening. The largest specimens may display a slight fold in the brachial valve and a corresponding depression (sulcus) in the pedicle valve. The shell of *Cancellothyris* has an external ornament of growth lines and numerous fine ribs.

HABITAT *Cancellothyris* was attached to hard surfaces by a stout pedicle. It is commonly found in clusters, especially in shallow-water, high-energy environments such as rocky shorelines. It also inhabited crevices and fissures.

large pedicle foramen

massive umbo

brachial valve

numerous fine ribs

Cancellothyris platys Brunton & Hillier; Alexander Bay Formation; Late Pliocene; South Africa.

Typical length ¾ in (2 cm)

slight fold

| Range: Miocene–Recent | Distribution: S. Africa, Australasia | Occurrence: ◉◉ |

| Group: TEREBRATULIDA | Subgroup: TEREBRATULIDAE | Informal name: Lamp shell |

Terebratula

The shell of this large terebratulid is elongate to elliptical in outline, with convex valves. There is a short, massive, and slightly curved-in umbo and a large, well-developed pedicle opening. Internally, the teeth and sockets are prominent, with a short middle septum on the brachial valve. At the front, the junction line is straight, with no fold or depression. Both valves are ornamented by fine growth lines, which become prominent near the shell margins.

HABITAT This large lamp shell had a thick, fleshy pedicle that branched into rootlets, attaching to shelly particles in unconsolidated sediments.

pedicle foramen

well-developed pedicle collar

junction line

growth lines

Terebratula maxima Charlesworth; Coralline Crag; Late Pliocene; UK.

Typical length 2½ in (6 cm)

| Range: Miocene–Pliocene | Distribution: Europe | Occurrence: ◉◉ |

BIVALVES

BIVALVES ARE MOLLUSKS in which the shells are made up of two valves connected by a ligament of organic material, rarely preserved in fossils. The valves are articulated by a hinge, usually with interlocking teeth. In most cases, the valves are closed by two main muscles whose points of attachment to the shell are marked by distinct scars. Bivalves feed by filtering particles from the water through siphons. While they possess a foot, they have limited mobility. Many burrow in sediment or bore into stone or wood. Others cement themselves to submerged objects or attach themselves with a byssus of organic threads. Bivalves are classified by their hinges.

Group: NUCULOIDA	Subgroup: NUCULANIDAE	Informal name: Nut clam

Nuculana

Lens-shaped in outline, the small shell has its beak positioned near the rounded anterior. The posterior ends in a curved, ridged rostrum. The valves are ornamented with concentric ribs. Two rows of interlocking hinge teeth are separated by a ligament pit below the beak.

HABITAT *Nuculana* burrows in mud and sand at a wide range of depths and temperatures.
REMARK Unlike other groups within the Nuculoida, there are no layers of nacre in the shell.

Nuculana marieana (Aldrich); Woods Bluff Beds; Early Eocene; US.

Joined valves

Interior

ribs

ligament pit

ridged rostrum

Exterior

Typical length ¾ in (2 cm)

Range: Triassic–Recent	Distribution: Worldwide	Occurrence: ◉◉◉

Group: PRAECARDIIDA	Subgroup: PRAECARDIIDAE	Informal name: Dog cockle

Slava

limestone matrix

Slava interrupta (J. de C. Sowerby); Silurian; Czech Republic.

The medium-sized shell is thin and convex. Below the beak in each valve is a triangular ledge with diverging grooves but no hinge teeth. The valves are ornamented with radiating ribs and spaced, concentric folds.

HABITAT It is believed that *Slava* lived just below the sediment in moderately deep waters.

concentric folds

radial ribs

Typical length ⅝ in (1.5 cm)

Range: Late Silurian–Devonian	Distribution: Europe, N. America	Occurrence: ◉◉◉

| Group: ARCIIDA | Subgroup: CUCULLAEIDAE | Informal name: Hooded ark shell |

Idonearca

The tumid shell has a straight hinge and a prominent beak. On the outside of each valve is a weak, radial ornament. Inside each valve, there is a row of small, vertical teeth, which become long and horizontal at both ends of the hinge.

HABITAT This mollusk burrows in sandy substrates. It is often associated with cold-water molluskan faunas.

REMARK *Idonearca* was widespread in the Late Cretaceous and Paleocene. A relict species survives in the West Pacific.

Idonearca vulgaris (Morton); Ripley Formation; Late Cretaceous; US.

beak

ornament

ligament area

end teeth turned outward

smooth margin

Typical length 1½ in (4 cm)

| Range: Early Jurassic–Recent | Distribution: Worldwide | Occurrence: ◉◉ |

| Group: ARCIIDA | Subgroup: GLYCYMERIDIDAE | Informal name: Bitter sweet |

Glycymeris

Within this genus, there are medium- to large-sized thick shells. The ornament of fine, radiating ribs crossed by growth lines weakens on the adult shell. The teeth are obliquely aligned outward from the middle of the hinge and are concentrated into lens-shaped rows on each side. The ligament area is triangular and grooved.

HABITAT *Glycymeris* burrows in sand and mud in shallow waters.

Glycymeris brevirostris (J. de C. Sowerby); London Clay; Early Eocene; UK.

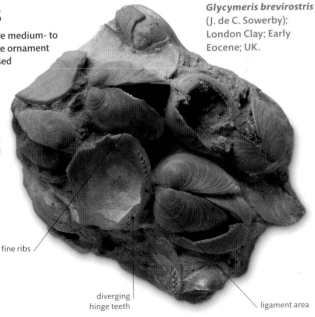

Typical length 1¼ in (3 cm)

fine ribs

diverging hinge teeth

ligament area

| Range: Early Cretaceous–Recent | Distribution: Worldwide | Occurrence: ◉◉◉◉◉ |

Group: MYTILIIDA	Subgroup: MYTILIDAE	Informal name: Horse mussel

Modiolus

The shell is elongate, thin, and tumid, with a beak very near the front. There is a straight hinge with a shore ligament, and the posterior margin is raised into an angular flange. The ribs are typically smooth but often have close diverging threads over the whole surface. The outer shell is brittle; the inside is lined with a layer of nacre.

HABITAT This bivalve usually lives gregariously in shallow water, attached to a solid object.

beak

Modiolus ligeriensis (d'Orbigny); Greensand; Late Cretaceous; France.

fine diverging ribs

long hinge

angular flange

Typical length 3 in (7.5 cm)

Range: Jurassic–Recent	Distribution: Worldwide	Occurrence: ●●

Group: PTERIOIDA	Subgroup: BAKEVELLIIDAE	Informal name: Tree oyster

Gervillaria

The moderately large, elongate shell has a small anterior ear and a large, triangular wing. The left valve is half-cylindrical and inflated; the right valve is flat, sometimes concave. Along the hinge line, there are several rectangular ligament pits, and below these a row of oblique teeth.

anterior ear

triangular wing with right-angled corner

Left valve

HABITAT This animal lived in warm, shallow seas, attaching itself to gravel or shell debris.

Gervillaria alaeformis (J. Sowerby); Lower Greensand; Early Cretaceous; UK.

Typical length 3⅛ in (8 cm)

Range: Jurassic–Cretaceous	Distribution: Europe	Occurrence: ●●●

Group: PTERIOIDA	Subgroup: PINNIDAE	Informal name: Fan mussel

Pinna

This large, wedge-shaped bivalve has weak, radial ribs running the length of the shell. In the living creature, the brittle calcitic valves are joined by a narrow ligament that runs along the entire hinge margin. Inside each valve is a pearly area, divided into two lobes by a distinct furrow, which shows as a slight ridge on the outside.

HABITAT *Pinna* lives in groups, with its pointed anterior end buried in the sediment.
REMARK When sediment fills the space between a pair of valves, it may harden into a siltstone nodule, preserving the fossil shell.

Pinna hartmanni
Zieten; Lower Lias;
Early Jurassic; UK.

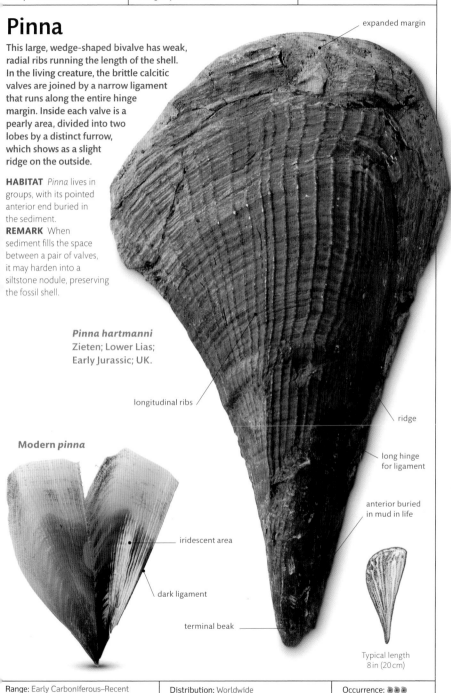

expanded margin

longitudinal ribs

ridge

long hinge for ligament

anterior buried in mud in life

terminal beak

Typical length 8in (20cm)

Modern *pinna*

iridescent area

dark ligament

Range: Early Carboniferous–Recent	Distribution: Worldwide	Occurrence: ◉◉◉

Group: MYALINIDA	Subgroup: INOCERAMIDAE	Informal name: Inoceramid oyster

Volviceramus

The shell of *Volviceramus* has an outer layer of calcite and an iridescent interior. Its straight hinge has narrow ligament pits. The left valve is large, thick, and spirally coiled, and the right valve sits within it. Concentric ridges are found on both valves, but they are stronger on the right valve.

HABITAT The growth and habits of this oyster are believed to have been similar to those of the genus *Gryphaea* (see p.101).

ligament pits on hinge

right valve

Typical length
12 in (30 cm)

concentric ridges

Volviceramus involutus
(J. de C. Sowerby);
Upper Chalk;
Late Cretaceous; UK.

Range: Late Cretaceous	Distribution: Europe, N. America	Occurrence: ⓐⓐⓐⓐⓐ

Group: PECTINIDA	Subgroup: AVICULOPECTINIDAE	Informal name: Scallop

Aviculopecten

The small shell is obliquely fan-shaped, with a short hinge. Both valves have the posterior ear enlarged into a pointed wing, while the anterior ear is set off by a byssal notch, much deeper in the right valve. The shell is ornamented with radial ribs. In one valve, the ribs increase by bifurcation, and in the other, by forming intermediary ribs.

HABITAT *Aviculopecten* was anchored to objects by a byssal thread.

byssal notch

Aviculopecten tenuicoullis (Dana);
Carboniferous;
Australia.

Typical length
1 in (2.5 cm)

posterior wing

Range: Carboniferous–Permian	Distribution: Worldwide	Occurrence: ⓐⓐⓐⓐ

Group: OSTREIDA	Subgroup: POSIDONIIDAE	Informal name: Paper mussel

Bositra

The roughly circular shell has concentric undulations, a short hinge, and poorly developed auricles.

HABITAT The lack of a clear byssal notch suggests that the animal might have been free-swimming, but it is more likely to have lived in colonies, attached to organic matter by a few byssal threads.

REMARK This genus was previously included in *Posidonia*.

Bositra radiata (Goldfuss); Upper Lias; Early Jurassic; UK.

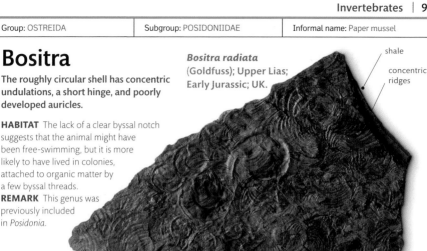

shale

concentric ridges

Crushed shells

pyrite formed by decayed organic material

Typical length 1¼ in (3 cm)

Range: Early Carboniferous–Late Jurassic	Distribution: Worldwide	Occurrence: ●●●●

Group: PECTINIDA	Subgroup: OXYTOMIDAE	Informal name: Scallop

Oxytoma

The posterior ear of this thin shell forms a sharply angular wing, while the front ear is very reduced. In the left valve, sharp, radiating ribs project as points on the margin, but the right valve is rounder and flatter, with a much finer ornament. The hinge is long and straight, with a small ligament pit near the middle.

HABITAT *Oxytoma* was a suspension feeder living in a variety of marine habitats and probably attached by a byssus to shell or gravel on the sea bed.

Oxytoma longicostata (Stuchbury); Lower Lias; Early Jurassic; UK.

sharp ribs on left valve

posterior ear

Typical length 1⅜ in (3.5 cm)

Range: Late Triassic–Late Cretaceous	Distribution: Worldwide	Occurrence: ●●●●

Group: PECTINIDA	Subgroup: PECTINOIDEA	Informal name: Scallop

Pecten

Convex right valve

The large, circular shell has a convex right valve and a flattened left valve; both valves are ornamented with broad, radiating ribs. The straight hinge has a central ligament pit, with diverging ridges on either side.

HABITAT Scallops like *Pecten* prefer clean sand in moderately shallow waters, where they rest on the bed in self-made depressions. When necessary, they swim by opening and closing their valves.

Pecten beudanti
Basterot;
Miocene;
France.

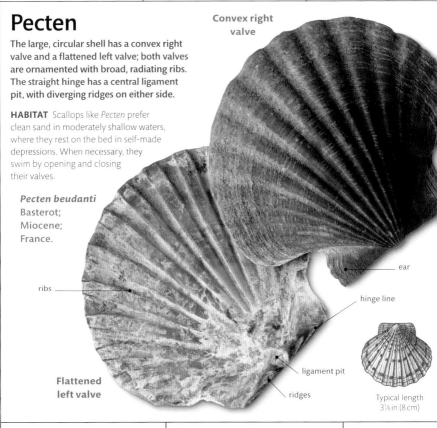

ribs

ear

hinge line

ligament pit

ridges

Flattened left valve

Typical length
3⅛ in (8 cm)

Range: Late Eocene–Recent	Distribution: Worldwide	Occurrence: ●●●●●

Group: PECTINIDA	Subgroup: PECTINOIDEA	Informal name: Scallop

Neithea

The shell is narrow and oval-shaped. Its main ribs have weaker ribs between them. The right valve is convex, overhanging the flattened left valve. There are denticles along the margin of the hinge.

HABITAT *Neithea* lived in sand, but it may have had the ability to swim.

main ribs

Neithea coquandi (Peron);
Late Cretaceous;
Tunisia.

Typical length
1½ in (4 cm)

Right valve

beak

Left valve

Range: Late Cretaceous	Distribution: Worldwide	Occurrence: ●●●●

Group: OSTREIDA	Subgroup: GRYPHAEIDAE	Informal name: Devil's Toenail

Gryphaea

The heavy shell is composed mostly of calcite. The concave right valve fits like a lid within the larger left valve, which has a narrow, inrolled beak. Internally, the single muscle scar is circular; the adult hinge has no teeth.

HABITAT *Gryphaea* lived on muddy sea beds, originally cemented to a small particle of rock by its tip.
REMARK The coiled shell, adapted for living in soft sediment, is popularly called the Devil's Toenail.

Gryphaea arcuata (Lamarck); Lower Lias; Early Jurassic; UK.

Two left valves

inrolled beak

coarse growth ridges

Typical length 2¾ in (7 cm)

Range: Late Triassic–Late Jurassic	Distribution: Worldwide	Occurrence: ◉◉◉◉◉

Group: OSTREIDA	Subgroup: GRYPHAEIDAE	Informal name: Oyster

Exogyra

Both valves grow spirally. The left valve is inflated and rather weakly ribbed, with the beak coiled off to one side. The flattened right valve is ornamented by concentric frills. The left valve hinge has two groups of ridges below the beak.

HABITAT *Exogyra* lived cemented to solid objects in warm seas.

Exogyra africana (Lamarck); Late Cretaceous; Algeria.

Typical length 2 in (5 cm)

Left valve, top view

concentric frills

Left valve, side view

beak

Right valve

Range: Late Cretaceous	Distribution: Worldwide	Occurrence: ◉◉◉◉

| Group: OSTREIDA | Subgroup: OSTREIDAE | Informal name: Oyster |

Ostrea

Ostrea has a rounded shell, with a large ligament pit on the hinge, below the beak. The single muscle scar is kidney-shaped.

HABITAT The oyster lives in marine or estuarine waters. The young often cement themselves to older mollusks, eventually forming a reef of variably shaped shells.

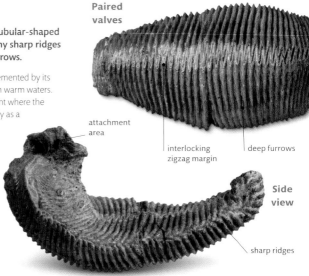

large ligament pit

growth ridges

Ostrea compressirostra (Say); Chesapeake Group; Miocene; US.

Typical length 4¾ in (12 cm)

| Range: Cretaceous–Recent | Distribution: Worldwide | Occurrence: ⊚⊚⊚⊚⊚ |

| Group: OSTREIDA | Subgroup: OSTREIDAE | Informal name: Cockscomb oyster |

Rastellum

The narrow valves of the tubular-shaped shell are composed of many sharp ridges divided by deep, sharp furrows.

HABITAT This animal lived cemented by its narrow beak to shell or coral in warm waters.
REMARK The vulnerable point where the valves meet is hidden, probably as a defense against predators.

Rastellum (Arctostrea) carinatum (Lamarck); Lower Chalk; Late Cretaceous; UK.

Paired valves

attachment area

interlocking zigzag margin

deep furrows

Side view

sharp ridges

Typical length 4 in (10 cm)

| Range: Middle Jurassic–Late Cretaceous | Distribution: Worldwide | Occurrence: ⊚⊚⊚⊚ |

Group: MODIOMORPHOIDA	Subgroup: MODIOMORPHIDAE	Informal name: False mussel

Whiteavesia

The triangularly ovate shell narrows toward the inconspicuous beak. The interior surface has ridges that appear on internal molds and probably reflect fine external ornament.

HABITAT A marine suspension feeder, it is likely that this genus was attached to the sea floor by byssus threads.
REMARK This early genus may have given rise to the Mytiloids (see p.96) and the Unionoids (see p.104), although it does not show the iridescence present in those two groups.

internal ridges

limestone matrix

Internal mold

Whiteavesia pholadiformis (Hall); Ordovician; Canada.

Typical length 2¾ in (7 cm)

Range: Middle–Late Ordovician	Distribution: N. America	Occurrence: ◑◉

Group: NUCULIDA	Subgroup: NUCULIDAE	Informal name: Nut clam

Nucula

The small ovoid shells are lined by shiny nacre on the inner surfaces. There are two rows of interlocking hinge teeth separated by the ligament scar. The inner lips of the shell are lightly serrated; the outer shell is ornamented with fine radial lines and concentric ribs.

HABITAT *Nucula* is a deposit feeder, living partially buried in muddy sand. It inhabits both coastal marine and deep-water environments.
HABITAT The interlocking hinge teeth and the tough, slippery nacre provide a defense against predation.

serrated lip

nacre

concentric growth lines

Nucula gracilenta S. V. Wood; Blackheath Formation; Early Eocene; UK.

hinge teeth

Typical length ⅝ in (1.5 cm)

Range: Cretaceous–Recent	Distribution: Worldwide	Occurrence: ◑◉

Group: UNIONOIDA	Subgroup: UNIONIDAE	Informal name: Freshwater mussel

Unio

Joined valves

The oblong to ovate shell is rather smooth, with a bluish or purplish pearl interior. The middle shell layer is prismatic and covered with a thin organic layer. There are two hinge teeth situated in front of the beak and two long teeth behind it.

HABITAT This animal lives in an upright position, partly embedded in the sands, silts, and gravels of lakes and rivers. Its larvae burrow under the scales of fish but drop off later, ensuring a wide dispersal.

ligament

beak

coarse growth lines

Unio menki (Koch & Dunker); Wealden Beds; Early Cretaceous; UK.

Typical length 3⅛ in (8 cm)

Range: Triassic–Recent	Distribution: Worldwide	Occurrence: ◉◉◉◉

Group: TRIGONIIDA	Subgroup: MYOPHORELLOIDAE	Informal name: Brooch clam

Mangyschlakella

Shells in sandstone

The thick valves of *Mangyschlakella* are semicircular, and each is divided into two parts: the main area, with obliquely concentric rows of tubercles, and the smaller posterior slope (the escutcheon), which is relatively smooth. Internally, the shells are iridescent. Two corrugated and diverging teeth, suspended below the beak of the right valve, fit into sockets in the left valve.

HABITAT This clam flourished in the shallow seas of the Mesozoic Era.

tubercle rows

eroded beak

Mangyschlakella elisae (Briart and Cornet); Bracquegnies Formation; Early Cretaceous; Belgium.

Typical length 2½ in (6 cm)

Range: Early Cretaceous	Distribution: Europe, western Asia	Occurrence: ◉◉

| Group: LUCINDA | Subgroup: LUCINIDAE | Informal name: Basket shell |

Fimbria

Fimbria has a tumid, ovate shell. The ornament is netted, with stronger concentric ridges at first, but then filelike, radial ribs that thicken at both ends of the valve. The whole shell margin is crenulated on the inside by a row of denticles. There are two main teeth below the beak and two lateral teeth in each valve.

HABITAT This suspension feeder lives in shallow, warm seas.

Fimbria subpectunculus (d'Orbigny); Calcaire Grossier; Middle Eocene; France.

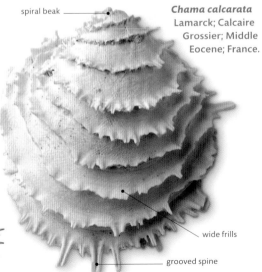

concentric ridges

Typical length 3⅛ in (8 cm)

netted ornament

| Range: Middle Jurassic–Recent | Distribution: Worldwide | Occurrence: ◉◉ |

| Group: VENEROIDEI | Subgroup: CHAMIDAE | Informal name: Jewel-box shell |

Chama

The left valve of this shell is thick and convex, cemented to the substrate by the spiral beak. The right valve may show a fine microsculpture. Both valves have concentric frills, bearing fragile, grooved spines. The hinge is degenerate, with one or two large, crudely shaped teeth that interlock.

HABITAT *Chama* inhabits sandy areas in warm, shallow seas, attached to rocks, corals, or shells.
REMARK The spines attract the growth of algae and other animals, which then act as a form of camouflage.

spiral beak

Chama calcarata Lamarck; Calcaire Grossier; Middle Eocene; France.

wide frills

grooved spine

Typical length 1⅜ in (3.5 cm)

| Range: Paleocene–Recent | Distribution: Tropics, Europe, N. America | Occurrence: ◉◉◉ |

Group: CARDITIDA	Subgroup: CARDITIDAE	Informal name: False cockle

Venericor

The large, thick, oval to triangular shell has a prominent beak. When very young, it has narrow, radial ribs, which may be armed with small tubercles. These soon become smooth, flat ribbons separated by shallow grooves. In each valve, the wide, triangular hinge plate bears three massive teeth. The inner shell margin is marked by a row of flat denticles.

HABITAT *Venericor* lived in warm, shallow waters, burrowing into the sand or silt.
REMARK Paired valves in life position are found in large numbers in some strata.

denticles

muscle scar

beak

hinge plate

Venericor planicosta
(Lamarck); Earnley
Formation; Middle
Eocene; UK.

broad, flat ribs

Typical length
3 in (7.5 cm)

Range: Paleocene–Eocene	Distribution: Europe, N. America	Occurrence: ◉◉◉◉

Group: CARDITIDA	Subgroup: ASTARTIDAE	Informal name: Astarte clam

Neocrassina

concentric ridges

The thick shell has regular concentric ridges. The hinge is thick, with three main teeth in the right valve and two in the left.

HABITAT *Neocrassina* burrows into mud, sand, or gravel offshore to considerable depths. It is most characteristic of the Boreal and Arctic regions.
REMARK Geologists regard it as a good indicator of cold conditions.

***Neocrassina*
shells in
limestone**

***Neocrassina
elegans***
(J. de C.
Sowerby);
Inferior Oolite;
Middle
Jurassic; UK.

Typical length
¾ in (2 cm)

Range: Jurassic–Recent	Distribution: Worldwide	Occurrence: ◉◉◉◉◉

| Group: CARDITIDA | Subgroup: CRASSATELLIDAE | Informal name: Clam |

Bathytormus

The solid rectangular shell has two V-shaped cardinal teeth in the left valve and one in the right, flanked by elongate lateral teeth. There are two muscle scars, which are joined by a roundly curved pallial line. The margins of the valves are finely crenulated. The young valves have strong concentric ridges. These fade out in some species but persist in others.

HABITAT These clams were common in warm, shallow seas, where they burrowed into soft substrates.
REMARK Some species of this genus evolved relatively rapidly.

hole made by predator

Bathytormus lamellosus (Lamarck); Calcaire Grossier; Middle Eocene; France.

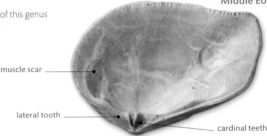

muscle scar

lateral tooth

cardinal teeth

Typical length
¾ in (2 cm)

| Range: Middle Cretaceous–Miocene | Distribution: Europe, N. America | Occurrence: ◉◉◉◉◉ |

| Group: CARDITIDA | Subgroup: CARDINIIDAE | Informal name: Clam |

Cardinia

The valves of the shell of *Cardinia* are ornamented with irregular, concentric bands. There is a triangular area, which is angled inward in front of the beak, deeply undercutting it. The lateral hinge teeth are thick and dominant.

HABITAT This appears to have been a marine genus, burrowing into sands, silts, and muds in warm, shallow waters.

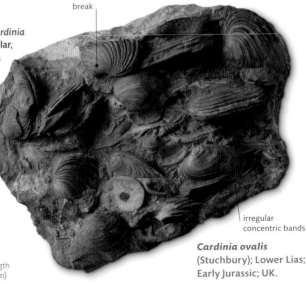

break

irregular concentric bands

Cardinia ovalis (Stuchbury); Lower Lias; Early Jurassic; UK.

Typical length
⅝ in (1.7 cm)

| Range: Late Triassic–Early Jurassic | Distribution: Worldwide | Occurrence: ◉◉ |

Group: CARDIIDA	Subgroup: CARDIIDAE	Informal name: Cockle

Acrosterigma

Smooth, radial ribs separated by grooves ornament the elliptical shell. The hinge plate is small and sharply curved. In the middle of the hinge of the left valve are two small, diverging teeth, which meet one narrow tooth in the right valve. The front end of the hinge is extended to form one or two lateral teeth. On the other side is a raised place (the nymph) where the ligament is attached. The inside edges of both valves are sharply furrowed.

HABITAT This creature burrows into the sands, silts, and muds of shallow tropical seas.

A range of shell sizes

well-developed lateral teeth

Typical length 3½ in (9 cm)

grooves separating smooth ribs

Acrosterigma dalli (Heilprin); Plio-Pleistocene; US.

Left valve exterior

Range: Late Oligocene–Recent	Distribution: Worldwide	Occurrence: ◉◉

Group: VENEROIDEI	Subgroup: ARCTICIDAE	Informal name: Iceland cockle

Arctica

The thick, smooth shell of *Arctica* is ovate to heart-shaped, with the beak turned toward the shorter anterior end. In each valve, the hinge has three teeth below the beak and a long lateral tooth behind. The hinge plate in front of the cardinal teeth is pitted and crenulated.

HABITAT This cockle rests on both firm and muddy sand, from the intertidal zone down to considerable depths.
REMARK The distribution of the genus indicates cold-water conditions. As fossils, they are regularly associated with other cold-water mollusks.

beak

Arctica umbonaria (Lamarck); Pliocene; Italy.

growth lines

Typical length 3½ in (9 cm)

Range: Late Cretaceous–Recent	Distribution: Europe, N. America	Occurrence: ◉◉◉◉

| Group: VENEROIDEI | Subgroup: CORBICULIDAE | Informal name: Marsh clam |

Polymesoda

Polymesoda has an oval shell, with weak concentric growth ridges and wrinkles. There are three cardinal teeth below the beak and strong, elongate lateral teeth. Some fossils still show radiating color bands.

HABITAT This clam lives in tidal marshes and brackish lagoons in tropical and subtropical regions.

REMARK European fossils differ from the American forms in details of the teeth and pallial line.

Slab with shells

concentric growth ridges

Polymesoda convexa (Brongniart); Bembridge Marls; Oligocene; UK.

Typical length 1½ in (4 cm)

| Range: Eocene–Recent | Distribution: Worldwide | Occurrence: ●●●●● |

| Group: VENEROIDEI | Subgroup: VENERIDAE | Informal name: Venus shell |

Lirophora

The ovate to triangular shell has its beak turned forward. The valves are ornamented with concentric cords and radial threads, and there are three diverging hinge teeth in each valve. Internally, the muscle scars are joined by the pallial line, which has a short embayment in front of the rear scar.

HABITAT This clam burrows below the sand in shallow seas.

Lirophora ceramota (Gardner); Chipola Formation; Early Miocene, US.

concentric cords

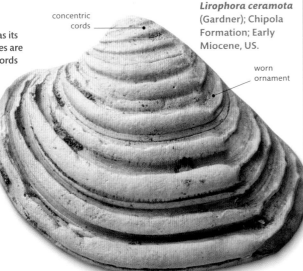

worn ornament

Typical length ⅝ in (1.6 cm)

| Range: Oligocene–Recent | Distribution: N. & S. America, New Zealand | Occurrence: ●● |

Group: MYOIDA	Subgroup: MYIDAE	Informal name: River clam

Potamomya

The shell is small and rather thin, varying from ovate to wedge-shaped. The surface is smooth, except where roughened by irregular growth lines. In the right valve, the ligament was attached to a central triangular projection; the rear part of this extends to form a narrow oblique tooth. The left valve hinge has a groove that receives this tooth.

HABITAT *Potamomya* was often abundant in nonmarine, silty sands and muds.

wedge-shaped shell

ovate shell

Typical length ⅝ in (1.5 cm)

Potamomya plana
(J. Sowerby); Headon Hill Formation; Late Eocene; UK.

Range: Late Eocene–Early Oligocene	Distribution: Europe	Occurrence: ●●

Group: HIATELLIDA	Subgroup: HIATELLIDAE	Informal name: Geoduck

Panopea

Coarse, irregular growth lines or, in some earlier species, regular concentric undulations are visible on the valves of the thin shell. There is one small central tooth in the hinge of each valve and a raised plate to which the ligament is attached.

HABITAT *Panopea* burrows in silt and mud.

broad and ridged beak

Panopea glycimeris
(Born); Pliocene; Italy.

concentric growth lines

flattened posterior end

Typical length 1⅜ in (3.5 cm)

Range: Early Cretaceous–Recent	Distribution: Worldwide	Occurrence: ●●●

Group: PHOLADIDA	Subgroup: PHOLADIDAE	Informal name: Shipworm

Teredina

Cross-section of log

Teredina was equipped with two small, inflated, triangular valves, cemented to a tube. When young, the two valves gaped widely both at front and rear, but later a callum was secreted to cover the anterior gape. As the animal grew, it enlarged its tube in stages.

Detail

HABITAT With the use of its valves, this animal bored its way into wood submerged in fresh to brackish water.

Teredina personata (Lamarck); London Clay Formation; Early Eocene; UK.

Overall length 2 in (5 cm)

animals bore toward the middle

Range: Late Cretaceous–Middle Miocene	Distribution: Europe	Occurrence: ◉◉

Group: PHOLADIDA	Subgroup: TEREDINIDAE	Informal name: Shipworm

Teredo

A fossil log in clay

The shipworm *Teredo* has serrated, convex, T-shaped valves. The hinge is a long, slender protuberance to which the muscle is attached.

HABITAT It lives gregariously in floating and sunken timber.

Teredo* cf. *antenautae J. Sowerby; London, Clay Formation; Early Eocene; UK.

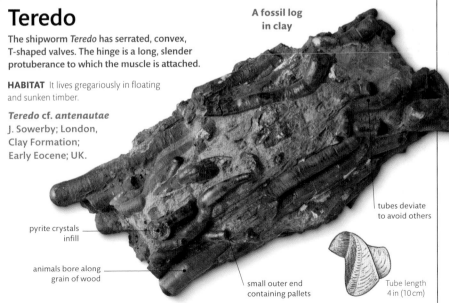

tubes deviate to avoid others

pyrite crystals infill

animals bore along grain of wood

small outer end containing pallets

Tube length 4 in (10 cm)

Range: Eocene–Recent	Distribution: Worldwide	Occurrence: ◉◉◉◉

Group: HIPPURITIDA	Subgroup: HIPPURITIDAE	Informal name: Rudist

Vaccinites

This genus belongs to an extraordinary group of complex bivalves known as rudists. One valve is conical in shape, enlarging rapidly from the base, where it is cemented to the substrate. The other valve is more or less flat, fitting like a small lid, with two long teeth descending from its inner surface and two bosses where the muscles were attached. The outside of the cone is corrugated by longitudinal ribs. Internally, the walls are folded, forming vertical pillars, a narrow tooth, and sockets to receive the teeth in the other valve. The structure of the shell allows water to pass through.

HABITAT *Vaccinites* lived near corals or on firm sand in warm seas. Although some species were solitary, others were gregarious and may have grouped together to form reefs.

Joined valves

lidlike upper valve

Vaccinites vesiculosus (Woodward); Late Cretaceous; Turkey.

lower valve with vertical ribs

attachment point

Typical length 4 in (10 cm)

Range: Late Cretaceous	Distribution: Europe, N.E. Africa, US, Asia	Occurrence: ◉◉◉◉◉

Group: CARDIIDIA	Subgroup: GRAMMYSIIDAE	Informal name: False razor shell

Solenomorpha

The fairly long, narrow shell of *Solenomorpha* is widest near the front, where the beak is situated, but then tapers gently behind. The paired valves, which have a slight gape at the posterior end, are smooth and have concentric growth lines.

HABITAT *Solenomorpha* burrowed in soft sediments.

tapering rear end

Solenomorpha minor (McCoy); Carboniferous Limestone; Early Carboniferous; UK.

Internal mold

Typical length ³⁄₄ in (1.8 cm)

Range: Early Devonian–Late Permian	Distribution: Worldwide	Occurrence: ◉◉◉

| Group: PHOLADOMYOIDA | Subgroup: PHOLADOMYIDAE | Informal name: Paper clam |

Pholadomya

This surface of the shell of *Pholadomya* is roughened by numerous minute pustules. The radial ribs, crossed by weaker concentric ridges, are strongest in midvalve. The hinge is thin, without teeth. The outer shell generally dissolves away, leaving a pearly film around the shell's internal mold. The interior is iridescent.

HABITAT *Pholadomya* inhabited shallow seas, buried in sand or mud. Surviving species live in deep water. Fossils can be found in life position, with the valves closed.

radial ribs

Typical length
1¹⁄₂ in (4 cm)

Pholadomya ambigua
(J. Sowerby); Middle Lias;
Early Jurassic; UK.

| Range: Late Triassic–Recent | Distribution: Worldwide | Occurrence: ◉◉◉◉◉ |

| Group: CARDIIDIA | Subgroup: SANGUNIOLITIDAE | Informal name: Clam |

Wilkingia

The thin shell has an iridescent interior. The beaks are placed close to the front of the shell, which has a long, rounded posterior end. The valves have strong, concentric undulations, usually stronger in the middle. The ends of the valves, which are almost smooth in some species, show radial rows of very fine pustules. The hinge line is without teeth but has a short ridge to hold the ligament.

HABITAT It burrowed in the soft sediments of shallow seas.

muscle scar

strongly ridged area

Typical length
1¹⁄₄ in (3 cm)

Wilkingia regularis (King *in*
de Verneuil); Great Limestone;
Early Carboniferous; UK.

| Range: Early Carboniferous–Permian | Distribution: Worldwide | Occurrence: ◉◉◉ |

SCAPHOPODS AND CHITONS

THE SCAPHOPOD SHELL is a tapering tube, open at both ends, with concentric ornament or longitudinal ribs. The larger end contains the animal's head and foot and is buried in the sediment; the smaller end is often notched and is regularly discarded as the animal grows. Chitons have a powerful foot for gripping hard surfaces, covered by a shell of eight overlapping valves held in place by a scaly or spiny girdle. The valves contain secretory and sense organs, which reach the surface as small pores.

Group: DENTALIIDA	Subgroup: DENTALIIDAE	Informal name: Tusk shell

Dentalium

The narrow, tusklike, tubular shell has strong, sharply angular, raised ribs running along its length. At the smaller, posterior end, the curvature is strong, but this tends to straighten with growth. The foot and tentacled head emerge from the larger, anterior end of the shell.

HABITAT *Dentalium* lives on the sea floor at a range of depths and temperatures.

strong, regular ribs

tube widens at the anterior end

Typical length 2 in (5 cm)

Dentalium sexangulum Gmelin *in* Linnaeus; Pliocene; Italy.

Range: Middle Triassic–Recent	Distribution: Worldwide	Occurrence: ⦿⦿⦿⦿⦿

Group: NEOLORICATA	Subgroup: LEPIDOPLEURIDAE	Informal name: Chiton

Helminthochiton

The large shell is made up of a narrow line of eight squarish valves, ornamented with granules and ridges. On the tail valve and median valves are two small flanges, embedded in the girdle in life and overlapping one another. The head valve is smaller, semicircular, and without the flanges. The whole shell is angled along the back, forming a blunt ridge from head to tail.

HABITAT *Helminthochiton* lived in warm seas on algae-covered debris.

Helminthochiton turnacianus (de Ryckholt); Early Carboniferous; Belgium.

strongly ridged median valve

Valves

tail valve

flange

ornament of ridges

Typical length 2³/₄ in (7 cm)

Range: Early Ordovician–Carboniferous	Distribution: Europe, N. America	Occurrence: ⦿

GASTROPODS

GASTROPODS ARE the largest, most successful class of mollusks and have been able to exploit a wide variety of marine, freshwater, and land habitats. They have a head with eyes and a mouth, a flattened foot for crawling, and viscera that are generally coiled and carried in a spiral shell. A few groups cease the coiling at some stage, and some have abandoned the shell altogether. Shells are composed of calcium carbonate, usually in the form of aragonite, but sometimes in layers of calcite. An inner nacreous layer is characteristic of some more early forms, but the organic outer coat of many living gastropods is usually lost in fossils. Some families have an inhalant siphon, emerging through a channel in the aperture.

Group: VETIGASTROPODA	Subgroup: BELLEROPHONTIDAE	Informal name: Sea snail

Bellerophon

The shell is coiled in a single plane, widening rapidly and flared at the aperture. A deep notch in the lip forms a ridge around the midline.

HABITAT This marine snail probably fed on vegetable matter.

Typical length
3 in (7.5 cm)

Bellerophon sp.;
Carboniferous
Limestone;
Carboniferous;
Belgium.

trumpetlike,
notched aperture

ribbed or
tubercular
rows

notches
forming central
ridged band

Top view

Range: Silurian–Early Triassic	Distribution: Worldwide	Occurrence: ⊕⊕⊕⊕

Group: VETIGASTROPODA	Subgroup: EUOMPHALIDAE	Informal name: Sea snail

Euomphalus

The disk-shaped, coiled shell has a slightly raised spire. The whorls are stepped, owing to a narrow notch in the growth lines, forming a sharp ridge. Successive internal walls seal off early whorls.

HABITAT It lived on marine vegetation.

Euomphalus pentangulus
(J. Sowerby); Carboniferous
Limestone; Carboniferous;
Ireland.

Typical length 2½ in (6 cm)

sharp ridge

Top view   **Cross-section**

Range: Silurian–Middle Permian	Distribution: Worldwide	Occurrence: ⊕⊕⊕

Group: VETIGASTROPODA	Subgroup: PLEUROTOMARIIDAE	Informal name: Slit shell

Pleurotomaria

Broadly conical, the shell has convex whorls separated by well-marked sutures. The outer layer is calcitic; the interior is nacreous. In the upper part of the lip is a deep, narrow notch. As the shell grew, the successive positions of this notch formed a groove (selenizone) around the shell, which in life served as an exhalant channel. The shell is ornamented with fine, spiral threads crossed by oblique lines, often thickened to form riblets on the shoulder of the whorl.

HABITAT This genus lived in deep waters worldwide.

conical spire

whorls with rounded shoulders

Typical length 2½ in (6 cm)

Pleurotomaria deshayesii (Deslongchamps); Middle Lias; Early Jurassic; France.

groove formed by lip notch

fine, spiral threads

Range: Early Jurassic–Early Cretaceous	Distribution: Worldwide	Occurrence: ◉◉◉

Group: VETIGASTROPODA	Subgroup: FISSURELLIDAE	Informal name: Keyhole limpet

Diodora

The low, conical shell has an oval margin. On the anterior side of the apex is a keyhole-shaped opening used by the animal as an exhalant channel. The shell is ornamented by radiating ribs, alternately thick and thin, crossed by growth lines.

HABITAT *Diodora* lives in shallow water, feeding on sponges and vegetable detritus.

exhalant opening

Shell exterior

Typical length 2 in (5 cm)

narrower anterior end

Diodora floridana Gardner; Pliocene; US.

alternating thick and thin ribs

Range: Late Cretaceous–Recent	Distribution: Worldwide	Occurrence: ◉◉◉

| Group: VETIGASTROPODA | Subgroup: SYMMETROCAPULIDAE | Informal name: Cap shell |

Symmetrocapulus

This limpet had a large, cap-shaped shell with an oval margin, ornamented with numerous narrow, radiating ribs interrupted by irregular, concentric folds. The first two small whorls are smooth, turned forward, and dextrally coiled, a little to the left of the shell's highest point.

HABITAT This marine limpet was adapted for grazing on rock surfaces.

beak

coiled apex, now eroded

Symmetrocapulus rugosus
(J. de C. Sowerby); Inferior
Oolite; Middle Jurassic; France.

narrow ribs

short anterior slope

Typical length
1¼ in (3.5 cm)

| Range: Jurassic–Eocene | Distribution: Europe, N. America | Occurrence: ◉ |

| Group: VETIGASTROPODA | Subgroup: PLATYCERATIDAE | Informal name: Sea snail |

Platyceras

The first few whorls of the shell are loosely coiled but rapidly enlarging, so that the last whorl is inflated and more or less cap-shaped. Sinuous growth lines produce prominent ridges and furrows that indent the shell margin. These had no functional significance but followed the shape of the object on which the animal lived—and so they vary between individuals.

HABITAT Commensal with crinoids, *Platyceras* shells are occasionally preserved in their life position on the host.

regularly coiled apical whorls

Platyceras haliotis
(J. de C. Sowerby);
Wenlock Limestone;
Silurian; UK.

margin follows shape of substrate

inflated body whorl, with ridges

Typical length
¾ in (2 cm)

| Range: Silurian–Early Carboniferous | Distribution: Worldwide | Occurrence: ◉◉◉◉ |

Group: VETIGASTROPODA	Subgroup: TROCHIDAE	Informal name: Top shell

Calliostoma

Shaped like a regular cone with flattened sides, the thinnish shell has shallow sutures separating the whorls. The whorls are ornamented with a number of fine, spiral cords, often bearing numerous round granules. These cords are continued on the flat base, which has no umbilicus. The oblique lip partly conceals the aperture.

HABITAT Various species inhabit rocky shores.

Calliostoma nodulosum (Solander *in* Brander); Barton Clay Formation; Middle Eocene; UK.

Typical length 1¼ in (3 cm)

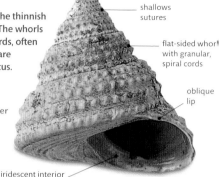

shallows sutures

flat-sided whorl with granular, spiral cords

oblique lip

iridescent interior

Range: Early Cretaceous–Recent	Distribution: Worldwide	Occurrence: ⦿⦿⦿

Group: VETIGASTROPODA	Subgroup: NERITIDAE	Informal name: Nerite

Velates

This nerite had a thick, heavy, oval shell with a low, conical spire. The early whorls are regularly coiled and visible at the apex, in the right posterior quarter of the shell. A layer of polished enamel over the spire obscures the spiral coiling and forms a thick pad over the base, ending in a toothed shelf halfway across the aperture. A separate shelly place (operculum) closed the aperture in life.

HABITAT *Velates* lived on sand in shallow waters, its shell's weight providing stability in strong currents.

partly obscured spiral apex

smooth, folded layer of polished enamel

thickened pad

narrow, semi-circular aperture

shelf with row of denticles

outer lip

Velates perversus (Gmelin); Early Eocene; France.

Typical length 1¼ in (3 cm)

Range: Early–Middle Eocene	Distribution: Worldwide	Occurrence: ⦿⦿

| Group: CAENOGASTROPODA | Subgroup: VIVIPARIDAE | Informal name: Pond snail |

Viviparus

The small- to medium-sized, thin, brittle shell has evenly rounded whorls, which are generally convex and smooth and crossed by faint growth lines. The lip is vertical in profile and the aperture oval with a continuous margin, slightly narrowed at the top. This margin almost hides the narrow umbilicus. In life, the aperture is closed by a flexible, horny plate.

HABITAT *Viviparus* lives in freshwater lakes, swamps, and slow rivers.

REMARK Because they can tolerate only the slightest degree of salinity, shells of *Viviparus* are an excellent indicator of nonmarine conditions in the fossil record.

Viviparus angulosus (J. Sowerby); Headon Hill Formation; Late Eocene; UK.

rounded whorls

umbilicus partly concealed by inner lip

edge of aperture

Typical length 1 in (2.5 cm)

| Range: Middle Jurassic–Recent | Distribution: Worldwide | Occurrence: ◉◉◉◉◉ |

| Group: CAENOGASTROPODA | Subgroup: COELOSTYLINIDAE | Informal name: Sea snail |

Bourguetia

The large and moderately thick shell of *Bourguetia* has convex whorls separated by well-marked sutures. Its extended spire appears slightly turreted and is a little longer than the final whorl. The upper part of the shell is ornamented with shallow grooves, dividing the surface into narrow, flat, spiral ribbons. On the basal part of the shell, these become thicker, more prominent cords. The aperture is oval with a continuous margin.

HABITAT Despite a superficial similarity to *Viviparus* (see above), *Bourguetia* was a fully marine snail that lived among coral and sponge reefs in warm and shallow seas.

regularly increasing whorls

moderately long spire with clearly indented suture

narrow, flat, spiral ribbons

Bourguetia saemanni (Oppel); Osmington Oolite; Middle Jurassic; UK.

more prominent cords on base

Typical length 4½ in (11 cm)

| Range: Middle Triassic–Late Jurassic | Distribution: Europe, New Zealand | Occurrence: ◉◉ |

| Group: CAENOGASTROPODA | Subgroup: THIARIDAE | Informal name: Marsh creeper |

Brotia

Brotia has a narrow, elongate, and turreted shell. When no longer in use, juvenile whorls are sealed off with a plug and then discarded or eroded away. The variable ornament usually consists of a few sharp, spiral cords broken up by the low, oblique ribs. The uppermost cord has the most prominent denticles and sets off the smooth concave shoulder. There is a broad, V-shaped embayment high on the lip and a weak channel at the base of the aperture.

HABITAT *Brotia* feeds on detritus in the mud of warm, brackish lagoons.

Brotia melanoides
(J. de C. Sowerby);
Woolwich Beds;
Paleocene; UK.

shelly plug
in apex

sharp, oblique
ribs

denticulate
shoulder cord

V-shaped growth lines showing
former positions of lip

broad, shallow
channel

Typical length 1½ in (4 cm)

| Range: Paleocene–Recent | Distribution: Europe, Asia | Occurrence: ●●● |

| Group: CAENOGASTROPODA | Subgroup: TURRITELLIDAE | Informal name: Screw turret |

Turritella

The shell of *Turritella* is a narrow, elongate cone with flattened sides. Whorls increase in size slowly, each one slightly overhanging the next. The tip of the spire is often sealed off and discarded, as in *Brotia*. The few distinct spiral cords on the early whorls increase in number as the shell grows. The aperture is nearly round, with a thin lip, often showing signs of breakages.

HABITAT *Turritella* is a suspension feeder, usually abundant in shallow off-shore waters.

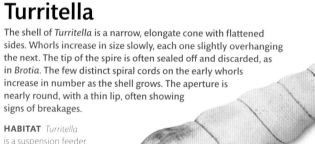

plugged apex

Turritella terebralis
Lamarck; Coquillat
de Leognan
Formation; Early
Miocene; France

thin lip

round aperture with no
channel at base

Typical length 2½ in (6 cm)

| Range: Late Eocene–Recent | Distribution: Worldwide | Occurrence: ●●●●● |

Group: CAENOGASTROPODA	Subgroup: CAMPANILIDAE	Informal name: Giant cerith

Campanile

A giant among gastropods, the shell of *Campanile* can reach a length of 24 in (60 cm), and if this could be uncoiled, it would make a tube over 10 ft (3 m) long. There may be more than 30 whorls constructed in the animal's lifetime, but the earlier ones are back-filled with shell once they have been evacuated and are periodically lost by abrasion, leaving the apex sealed. Early whorls are turreted by an upper row of tubercles, with finer cords below. On the last few whorls, the ornament deteriorates, leaving a single row of blunt knobs. The large, flared aperture is developed only on fully adult shells. The central pillar (columella) is strengthened by two encircling ridges that are obsolete at the aperture but can be seen clearly when the shell is broken. A calcified outer layer is covered with rows of small pits.

HABITAT *Campanile* feeds on algae on sandy bottoms in very shallow, warm seas. The tip of the heavy shell trails behind it as it crawls, eventually wearing flat on one side.
REMARK Only a single species survives today, in the shallow coastal waters off Australia.

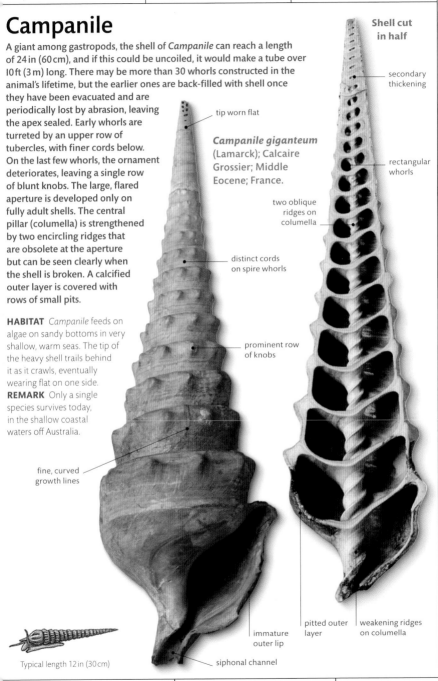

Shell cut in half

secondary thickening

Campanile giganteum (Lamarck); Calcaire Grossier; Middle Eocene; France.

tip worn flat

rectangular whorls

two oblique ridges on columella

distinct cords on spire whorls

prominent row of knobs

fine, curved growth lines

immature outer lip

siphonal channel

pitted outer layer

weakening ridges on columella

Typical length 12 in (30 cm)

Range: Late Cretaceous–Recent	Distribution: Worldwide	Occurrence: ◉◉◉

Group: CAENOGASTROPODA	Subgroup: APORRHAIDAE	Informal name: Pelican's foot

Tessarolax

The small, biconical, ridged shell of *Tessarolax* was much enlarged by having the extremities drawn out into four slender and delicate, curved spines. The bases of these are connected by an outgrowth of the lip, like the web of a duck's foot.

HABITAT This creature was probably a detritus feeder, living on muddy sand in moderately deep water.

REMARKS The spines may have helped prevent the shell sinking into the substrate.

spine joined to spire

curved basal spine

lip enlarged with two labial spines

Tessarolax fittoni (Forbes); Lower Greensand; Early Cretaceous; UK.

Typical length 2½ in (6 cm)

Range: Cretaceous	Distribution: Worldwide	Occurrence: ◉

Group: CAENOGASTROPODA	Subgroup: STRUTHIOLARIIDAE	Informal name: Ostrich foot

Struthiolaria

This sturdy, medium-sized mollusk has a proportionately large body whorl and a turreted spire. Distinct sinuous lines mark the positions of the lip during growth. When adult, both the lip and the inner margins of the aperture are thickened with enamel, which spreads to form a pad. Spiral cords often ornament the whole shell, with a series of ribs or nodes on the whorl shoulder. Some species, however, are smooth and polished.

HABITAT This genus feeds on algal detritus in shallow waters.

stepped spire whorls with deep sutures

growth lines following lip profile

aperture enameled and thickened into a pad

Struthiolaria ameghinoi (von Ihering); Santa Cruz Formation; Early Miocene; Argentina.

Typical length 1½ in (4 cm)

Range: Paleocene–Recent	Distribution: Australasia, S. America	Occurrence: ◉◉◉

Group: CAENOGASTROPODA	Subgroup: STROMBIDAE	Informal name: Beak shell

Rimella

The elongate shell of *Rimella* had gently convex whorls, ornamented with numerous sharp ribs. The spiral sculpture consists of close cords, largely obsolete in some species. The aperture is narrow and lens-shaped, and the last whorl tapers down to a narrow rostrum, separated from the lip by a distinct notch. When adult, a long, narrow channel was secreted from the upper corner of the aperture to the apical area, where it was usually bent around. As this channel was too narrow to accommodate the siphon, its function is unclear.

HABITAT *Rimella* was probably a herbivorous snail that lived in the muds and sands of warm, shallow seas.

Rimella fissurella (Linnaeus); Calcaire Grossier; Middle Eocene; France.

tip of channel bent near apex

convex whorls

notch in base of lip

inner channel connects with aperture

reflected lip

Typical length 1 in (2.5 cm)

Range: Paleocene-Oligocene	Distribution: Worldwide	Occurrence: ◉◉◉◉

Group: CAENOGASTROPODA	Subgroup: HIPPONICIDAE	Informal name: Sorting hat shell

Hipponix

The small- to medium-sized, caplike shell tends to be rather variable in shape. The juvenile shell is smooth and coiled slightly to the right of the apex but is soon worn away, leaving a blunt point. The adult shell expands rapidly to form a low cone with an extended beak, which often overhangs the rear shell margin. It is ornamented with close, fine, radiating ridges.

HABITAT *Hipponix* lives on other shells and corals in warm, shallow seas, feeding on particles in suspension.

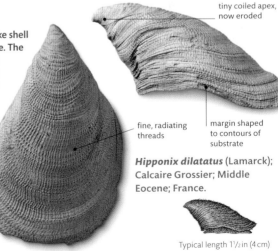

tiny coiled apex, now eroded

fine, radiating threads

margin shaped to contours of substrate

Hipponix dilatatus (Lamarck); Calcaire Grossier; Middle Eocene; France.

Typical length 1½ in (4 cm)

Range: Late Cretaceous–Recent	Distribution: Worldwide	Occurrence: ◉◉◉

Group: CAENOGASTROPODA	Subgroup: CALYPTRAEIDAE	Informal name: Slipper limpet

Crepidula

Beginning with one simple juvenile whorl, *Crepidula* rapidly expands into an elongate-oval body whorl, smooth in some species but ridged or spiny in others. On the underside, the wide aperture is half covered by a thin, polished shelf (septum), which protects the viscera.

HABITAT The animal is a filter feeder in shallow seas. Its sedentary habit is indicated by the irregular, wavy shell margin, which conforms to the shape of the substrate.

growth line showing earlier margin

Crepidula falconeri Newton; Middle Eocene; Nigeria.

Typical length ¾ in (2 cm)

simple, one-turn, spiral apex

irregularly wavy margin matches substrate

Range: Late Cretaceous–Recent	Distribution: Worldwide	Occurrence: ●●●

Group: CAENOGASTROPODA	Subgroup: XENOPHORIDAE	Informal name: Carrier shell

Xenophora

This has a moderately low, conical shell with shallow sutures, ornamented with oblique, interrupted ridges. The flat base with its narrow aperture is overhung by a thin, wavy peripheral margin.

HABITAT Hidden beneath its shell, the animal browses on micro-organisms in the silty mud of fairly deep waters.

REMARK *Xenophora* has acquired the habit of selecting bits of shell or other debris from the sea bed and cementing them to the lobes of the periphery, where they become incorporated.

Basal view

constricted aperture with concave lip

Xenophora crispa (König); Pliocene; Italy.

smaller debris attached when younger

periphery of earlier whorl

objects cemented to the margin support the shell

Side view

peripheral lobes

Typical length 1¼ in (3 cm)

Range: Late Cretaceous–Recent	Distribution: Worldwide	Occurrence: ●●●

Group: CAENOGASTROPODA	Subgroup: CYPRAEIDAE	Informal name: Egg cowrie

Umbilia

This cowrie has a large, globular shell with a flattened underside, narrowed at each end. The spire is a flat coil obscured by a shiny enamel coating, which covers the rest of the shell. When adult, the lip is inflated and wholly curled inward, narrowing the long aperture. This curves toward the top and is guarded by a line of strong, ridgelike teeth on the body whorl and by a row of finer denticles on the lip.

HABITAT Most cowries feed on detritus, but the habits of *Umbilia* are not well known.

outer lip curled inward

apical spout, bent sideways

flat spire, partly covered

aperture

small tubercles

globular body whorl, glazed with layer of enamel

lips lined with different-sized teeth

Typical length
1¼ in (3.5 cm)

Umbilia eximia (G. B. Sowerby);
Miocene; Australia.

Range: Middle Oligocene–Recent	Distribution: Australia	Occurrence: ◉

Group: CAENOGASTROPODA	Subgroup: NATICIDAE	Informal name: Moon shell

Natica

The smooth, globular shell has a large body whorl with a much smaller spire and an umbilicus with a central ridge. The outer lip is thin, and the aperture is closed in life by a right-fitting operculum.

HABITAT This gastropod lives at various depths in marine to slightly brackish waters. It is carnivorous, drilling bevel-edged holes in other mollusk shells and sucking out the contents.

Natica shell, apical view

Prey shell

feeding hole drilled in bivalve

Typical length
1¼ in (3 cm)

Natica stercusmuscarium (Gmelin); Pliocene; Italy.

Range: Paleocene–Recent	Distribution: Worldwide	Occurrence: ◉◉◉◉◉

| Group: CAENOGASTROPODA | Subgroup: CASSIDAE | Informal name: Helmet shell |

Phalium

Phalium has a strong, medium-to-large shell, with an inflated body whorl ornamented with both spiral and transverse threads and two to four spiral rows of tubercles. The uppermost of these is situated on the angulated shoulder of the whorl, giving rise to a low, flat-sided, conical spire. The reflected and thickened lip is corrugated inside, and a roughly ridged layer of enamel is spread around the aperture on to the body whorl. At the bottom of the aperture, and bent away from it, is the tubular siphonal channel.

HABITAT *Phalium* is an active predator, living on coral rubble or sand in warm waters.

Typical length
4 in (10 cm)

Phalium decussatum
(Linnaeus); Calcaire
Grossier; Middle
Eocene; France.

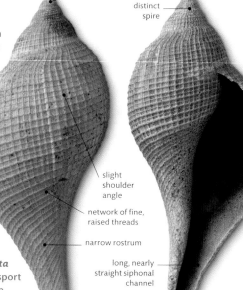

reflected lip

flat-sided spire

row of shoulder nodes

tubular siphonal channel

| Range: Paleocene–Recent | Distribution: Worldwide | Occurrence: ⬤⬤⬤⬤ |

| Group: CAENOGASTROPODA | Subgroup: FICIDAE | Informal name: Fig shell |

Ficopsis

Ficopsis had a thinnish shell tapering to a narrow rostrum. The raised spire, which is much shorter than the body whorl, has gently convex whorls, a narrow apex, and well-defined sutures. A latticework of fine, regular threads covers the whole shell. The body whorl is convex and varies from being evenly rounded to having a polygonal profile, caused by the accentuation of the slight shoulder and spiral ridges. The aperture is thin-lipped and ovate, narrowing into a long siphonal channel.

HABITAT This animal lived on sandy substrates in warm regions.

Typical length
2¹/₂ in (6 cm)

Ficopsis penita
(Conrad); Gosport
Sands; Middle
Eocene; US.

narrow apex

distinct spire

slight shoulder angle

network of fine, raised threads

narrow rostrum

long, nearly straight siphonal channel

| Range: Eocene | Distribution: N. America, Europe | Occurrence: ⬤⬤⬤ |

| Group: CAENOGASTROPODA | Subgroup: EPITONIIDAE | Informal name: Wentletrap |

Cirsotrema

Cirsotrema has an elongate calcitic shell, with convex whorls separated by deep sutures and an ornament of numerous raised and frilled ribs, each made up of a number of compressed laminae. In some species, they connect up from one whorl to the next. The last rib frames the small, circular aperture. A few weak spirals are visible between, and sometimes cross, the ribs, while the base of the shell is surrounded by a strong cord.

HABITAT This marine-dwelling animal is carnivorous, using its long proboscis to feed on sedentary creatures.

deep sutures

ribs made up of laminae

less prominent spiral ornament

basal cord

circular aperture with no siphonal channel

Cirsotrema lamellosum
(Brocchi); Pliocene; Italy.

Typical length
1¼ in (3 cm)

| Range: Paleocene–Recent | Distribution: Worldwide | Occurrence: ◉◉◉◉ |

| Group: CAENOGASTROPODA | Subgroup: MURICIDAE | Informal name: Comb shell |

Murexsul

Murexsul has a medium-sized shell with sharp ribs and a long rostrum. The whorls are sharply shouldered and the sutures are deep, giving rise to a turreted spire ornamented with closely packed spiral cords. These rise at intervals to form a rib, composed of a wall of fluted spines. The last rib surrounds the smooth, enameled aperture, slightly detached from the rest of the shell. The base of the aperture is suddenly narrowed into a siphonal channel, which is virtually closed to form a tube, and this rostrum is armed with a few more spines.

HABITAT *Murexsul* is a carnivore, inhabiting moderately deep waters.

turreted spire with deep sutures

thick, rounded, spiral ridges

rib made of row of tubular spines

spiny rostrum

Typical length
3½ in (9 cm)

Murexsul octogonus
(Quoy & Gaimard);
Late Pliocene;
New Zealand.

| Range: Miocene–Recent | Distribution: Australasia, W. Pacific | Occurrence: ◉◉ |

Group: CAENOGASTROPODA	Subgroup: MURICIDAE	Informal name: Sea snail

Ecphora

Ecphora is one of the best known and most striking fossil gastropods. The spire of its shell is relatively small and turreted, and the whorls are dominated by two prominent encircling keels that are often grooved along their length. One or two more keels appear on the base of the body whorl. The siphonal channel is sharply bent back from the aperture, and a chain of the spouts of previous channels makes a spiral around the rostrum.

HABITAT This creature was a carnivorous predator, living in shallow waters.

Typical length 4 in (10 cm)

turreted spire

simple sculpture of strong keels

suture

basal wall extension

tubular ends of former channels

Ecphora quadricostata (Say); Yorktown Formation; Pliocene; US.

Range: Late Oligocene–Pliocene	Distribution: N. America, Europe	Occurrence: ◉◉◉◉

Group: CAENOGASTROPODA	Subgroup: NEPTUNEIDAE	Informal name: Whelk

Neptunea

The shell of this large whelk has rounded whorls separated by deep sutures. Some species coil dextrally, as most gastropods do, but others are typically sinistral. A fairly long and stepped spire leads in to a convex body whorl. The large, ovate aperture has a rather short and broad siphonal channel. If present, ornament is restricted to spiral lines, cords, or keels, although coarse, transverse growth lines may occur.

HABITAT A cold-water carnivore, *Neptunea* preys on bivalves and other invertebrates in moderately to very deep seas.

sinistral coiling

Neptunea angulata Harmer; Red Crag; Late Pliocene; UK.

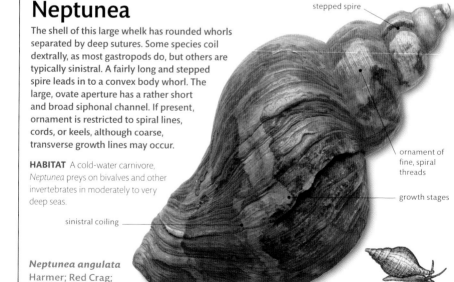

stepped spire

ornament of fine, spiral threads

growth stages

Typical length 3½ in (9 cm)

Range: Late Eocene–Recent	Distribution: Worldwide	Occurrence: ◉◉◉

Group: CAENOGASTROPODA	Subgroup: FASCIOLARIIDAE	Informal name: Sea snail

Clavilithes

Clavilithes had a medium to large, fusiform shell. The young shell begins as a smooth column of equal-sized whorls, developing low ribs crossed by spiral threads. The sculpture soon dies away, leaving only a few spiral lines, and the whorls become flat-sided. A sharp shelf at the top of the whorl gives the spire a stepped shape. The last whorl is usually cup-shaped and the aperture oval.

HABITAT Probably carnivorous, this gastropod was an inhabitant of sand, silt, and mud in moderately deep, warm waters.

Typical length 4¾ in (12 cm)

early sculpture

stepped spire with smoother whorls

shelf

Cross-section

ovate aperture with siphonal canal

Clavilithes pinus (Perry); Barton Clay Formation; Middle Eocene; UK.

body whorl narrows abruptly

Range: Paleocene–Pliocene	Distribution: Worldwide	Occurrence: ◉◉◉◉

Group: CAENOGASTROPODA	Subgroup: VOLUTIDAE	Informal name: Volute

Volutospina

The medium to large, biconic shell has an acutely pointed apex. Despite the shallow sutures, the spire has a turreted appearance caused by the short ribs, which end in sharp spines on the shoulder. The body whorl is ornamented with weak, flat, spiral ribbons crossed by numerous inconspicuous vertical growth lines. The rather narrow aperture tapers down to an open siphonal channel, and the outer lip is not thickened. There are several ridges on the columella, the lowest one being the strongest, and a layer of enamel spreads from the interior on to the body whorl.

HABITAT This creature inhabited sandy or muddy sea beds in warm waters.
REMARK Like other volutes, *Volutospina* was a fast-moving predator.

Volutospina luctator (Solander *in* Brander); Barton Clay Formation; Middle Eocene; UK.

narrow apex

low ribs with prominent spines

Typical length 2¾ in (7 cm)

rostrum with siphonal channel

Range: Late Cretaceous–Pliocene	Distribution: Worldwide	Occurrence: ◉◉◉◉◉

Group: CHAENOGASTROPODA	Subgroup: PSEUDOLIVIDAE	Informal name: False olive

Pseudoliva

The medium-sized, near-spherical shell of *Pseudoliva* has a depressed spire with a small, central, projecting apex. The ovate body whorl has a humped shoulder caused by enamel being secreted below the suture. Species are variably ornamented with nodes, ribs, or coarse growth lines. The aperture is wide, with a broad siphonal channel. Low down on the thin outer lip is a deep and narrow depression, which gives rise to a sharp groove around the shell.

HABITAT *Pseudoliva* lives on sand and silt in shallow waters.
REMARK One species survives in west Africa.

Pseudoliva laudunensis (Defrance); London Clay; Early Eocene; UK.

low spire

build-up of enamel below suture

apex

wide and shallow siphonal channel

groove ends in lip notch

sharply incised groove around base of whorl

Typical length 1¼ in (3 cm)

Range: Late Cretaceous–Recent	Distribution: Worldwide	Occurrence: ◉◉◉◉

Group: CHAENOGASTROPODA	Subgroup: CONIDAE	Informal name: Cone shell

Eoconus

Eoconus has a strong, medium-sized shell with a largely smooth, conical body whorl, strongly angled to form a narrow platform at the top. The spire is stepped and depressed, with a prominent central apex. A thin lip, parallel with the body, forms the straight, narrow aperture running the length of the whorl.

HABITAT These creatures live in warm waters on sand, silt, or hard bottoms at a variety of depths.
REMARK The Conidae are predatory; they hunt worms, mollusks, and even fish using teeth that are adapted to function as effective poison darts.

Eoconus sauridens (Conrad); Stone City Formation; Middle Eocene; US.

low, tightly coiled spire

platform at top of whorl

thin outer lip

narrow channel and aperture

Typical length 2½ in (6 cm)

Range: Late Cretaceous–Recent	Distribution: Worldwide	Occurrence: ◉◉◉◉

| Group: CHAENOGASTROPODA | Subgroup: CLAVATULIDAE | Informal name: Turrid |

Eosurcula

long spire

This turrid had an elongate, spindle-shaped shell with loosely coiled whorls. The spire is long and slender, with shallow but distinct sutures, and the whorl profile is weakly angled below the shoulder. The lens-shaped aperture gradually tapers into the siphonal channel, housed in a long, fine rostrum. Just below the top of the thin, curved lip is a characteristically deep, rounded sinus. The shell is ornamented with strong spiral threads, crossed by close, fine growth lines.

HABITAT Like other turrids, *Eosurcula* was a predator, living on sand in moderately deep waters.

strong, spiral threads below shoulder

sinus band on shoulder

anal sinus in lip

siphonal channel in long rostrum

Eosurcula moorei (Gabb); Stone City Formation; Middle Eocene; US.

Typical length 1¼ in (3 cm)

| Range: Eocene | Distribution: N. America | Occurrence: ●●● |

| Group: HETEROBRANCHIA | Subgroup: ARCHITECTONICIDAE | Informal name: Sundial shell |

Granosolarium

Basal view

Granosolarium has a low, conical shell on which the tiny initial whorls are smooth and slightly inturned and marked off by a collar. Beneath the sharply angled periphery, the base is almost flat. In the center, the wide, funnel-shaped umbilicus exposes the lower side of the whorls all the way to the apex. The whole shell is ornamented with granulated, spiral cords crossed by oblique growth lines.

HABITAT This mollusk lives in shallow to moderately deep waters, probably as a parasite of corals or other sedentary animals.

triangular aperture

sharply angled periphery

base with wide umbilicus

ornament of granulated spiral cords

Granosolarium elaboratum (Conrad); Gosport Sands; Middle Eocene; US.

Top view

Typical length 1½ in (4 cm)

| Range: Late Cretaceous–Recent | Distribution: Worldwide | Occurrence: ●●●●● |

Group: CEPHALASPIDEA	Subgroup: BULLIDAE	Informal name: Bubble shell

Bulla

Bulla has a rather thin, almost spherical shell, only one whorl of which is visible. The few tiny initial whorls coil sinistrally, but the shell then resumes dextral coiling. Each whorl wholly encloses the previous one, so there is no apparent spire. The aperture is at least as long as the shell, and an enamel layer strengthens the columella wall.

HABITAT Unlike some of its nearest relatives, this is a herbivorous animal living on sandy or muddy bottoms in shallow, warm waters.

REMARK An unusual feature of this gastropod is that its body is twice the size of the shell and is unable to withdraw into it.

> **Bulla ampulla**
> Linnaeus; Reef limestone;
> Pleistocene; Red Sea.

apical umbilicus

top of aperture reaches apex

near-spherical body whorl

enamel on columella wall

Typical length
³/₄ in (2 cm)

Range: Late Jurassic–Recent	Distribution: Worldwide	Occurrence: ◉◉◉

Group: HETEROBRANCHIA	Subgroup: NERINEIDAE	Informal name: Auger shell

Nerinea

Nerinea had a long, narrow shell, typically with concave sides to the whorls and accentuated keels just over the sutures. The apex, only rarely preserved, is a tiny, flat coil that stands at right angles to the axis of the main shell. A characteristic of the genus is a collection of fine ridges and processes on the walls of the small aperture; there may be up to seven of these, best seen in cross-sections.

HABITAT *Nerinea* is often associated with coral reef deposits.

> **Nerinea sp.;**
> Cretaceous; Israel.

concave whorls between spiral keels

columella ridges

columella ridge visible in aperture

internal ridge

Typical length
2¹/₂ in (6 cm)

Young shell in cross-section

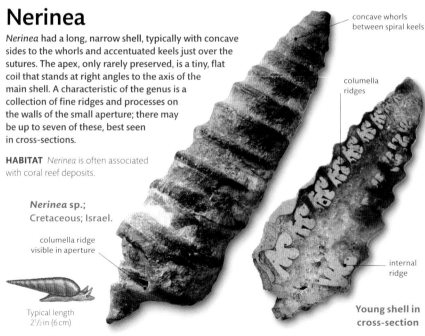

Range: Early Jurassic–Late Cretaceous	Distribution: Worldwide	Occurrence: ◉◉◉◉

Group: BASOMMATOPHORA	Subgroup: PLANORBIDAE	Informal name: Ramshorn snail

Australorbis

Rather thin and disk-shaped, the shells of this family are in fact sinistrally coiled, although they appear to be dextral. The inverted spire takes the form of a deep and widely open umbilicus containing close, convex whorls. Bluntly angled at the periphery, the base is flat and ornamented with occasional weak, spiral lines. Oblique growth lines cover the whole shell. The aperture is small and triangular/oval.

HABITAT Today, the family, Planorbidae, is known to live exclusively in fresh water, and the genus *Australorbis* feeds on vegetable matter in lakes, marshes, and rivers. It is fairly certain that this was always the case, so this fossil is a useful indicator of nonmarine conditions in the past.

flat base with faint ornament

Basal view

sinistral shell that appears to be built upside down

Top view

small, oval aperture

Australorbis euomphalus (J. Sowerby); Headon Hill Formation; Late Eocene; UK.

deeply sunk whorls

Typical length 1¼ in (3 cm)

Range: Late Cretaceous–Recent	Distribution: Europe, Asia, N. & S. America	Occurrence: ◉◉◉◉

Group: STYLOMMATOPHORA	Subgroup: CLAUSILIIDAE	Informal name: Land snail

Rillyarex

shallow sutures

The large shell is sinistrally coiled, unlike most gastropods. The elongate spire is narrowly arched in profile, with gently rounded sides, hardly indented by the shallow sutures and slightly convex whorls. The whole shell is ornamented with numerous oblique, transverse ridges. The ovate aperture has a continuous, slightly flared rim except at its upper junction.

HABITAT Although larger than its living relatives, *Rillyarex* had all the characteristics of an air-breathing terrestrial snail and was probably washed into the freshwater deposits in which it is found.

Rillyarex preecei Nordsieck; Bembridge Limestone; Late Eocene; UK.

ridged ornament

Typical length 2¾ in (7 cm)

aperture with rim

Range: Middle Eocene–Oligocene	Distribution: Europe	Occurrence: ◉◉

NAUTILOIDS

NAUTILOIDS ARE EARLY, marine cephalopods that possess a shell. They were most abundant in the Paleozoic Era, 400 million years ago; today, only a single genus survives—the pearly nautilus of the southwest Pacific Ocean. The shell is divided into a body chamber and many smaller chambers. The chambered part of the shell is known as the phragmocone. Nautiloids have heads with well-developed eyes and grasping tentacles. They swim by squirting water out of the body cavity.

Group: ORTHOCERIDA	Subgroup: ORTHOCERATIDAE	Informal name: Orthoceras

Orthoceras

The shell of *Orthoceras* is shaped like a slowly tapering cylindrical cone and is made up of closely spaced concavo-convex chambers, joined together by a centrally placed tube called the siphuncle.

HABITAT *Orthoceras* was an active swimmer, with the shell positioned horizontally in the water. It scavenged and predated on small animals.

separate chambers of phragmocone

Internal mold of phragmocone

convex posterior of chamber

Orthoceras sp.; MacDonnell Ranges; Middle Ordovician; Australia.

Typical length 6 in (15 cm)

Range: Middle Ordovician	Distribution: Worldwide	Occurrence: ⓐⓐⓐⓐⓐⓐ

Group: ORTHOCERIDA	Subgroup: CYRTOCERATIDAE	Informal name: Nautiloid

Cyrtoceras

The shell is strongly curved in an open hook shape and nearly circular in cross-section. The posterior part is made up of numerous closely spaced chambers, while the long living chamber makes up the anterior part of the shell.

HABITAT Experiments suggest that the animal lived head-down in the water, probably near the bottom of the sea. It was able to swim and alter its buoyancy.
REMARKS The specimen featured is an internal mold, from which the original shell has been completely lost.

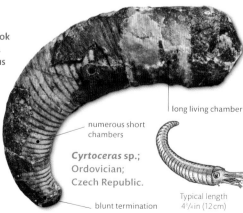

long living chamber

numerous short chambers

Cyrtoceras sp.; Ordovician; Czech Republic.

blunt termination

Typical length 4¾ in (12 cm)

Range: Ordovician	Distribution: Worldwide	Occurrence: ⓐⓐ

Group: ACTINOCERATIDA	Subgroup: HURONIIDAE	Informal name: Nautiloid

Huronia

Shells of this genus have a large, straight siphuncle, with long segments separated by strong constrictions. Complex canals are present. The central canal is very narrow.

HABITAT This genus lived in shallow seas as a scavenger or predator. It was able to swim and adjust its position in the water column by moving liquid along the siphuncle from chamber to chamber.
REMARK Only the robust siphuncle of this species is preserved in the specimen. The chambers and outer wall are missing.

limestone matrix

large segments of siphuncle

Huronia vertebralis
Stokes; Ordovician; Canada.

strong constrictions

Typical length 8 in (20 cm)

Range: Ordovician–Silurian	Distribution: N. America	Occurrence: ◉◉

Group: ORTHOCERIDA	Subgroup: ORTHOCERATIDAE	Informal name: Orthoceras

Orthoceras limestone

The late Silurian and early Devonian Orthoceras limestones of the Erfoud area in southeastern Morocco are well known for their spectacularly preserved orthocone cephalopods and goniatites. The stone is extensively quarried and turned into a variety of items—including crockery, paper weights, bathroom sinks, and huge wall plaques—that are sold worldwide. The orthocone assemblage comprises scores of different species; however, one of the most abundant is *Orthocycloceras fluminense*.

HABITAT Orthoceras limestone was deposited in a shallow, highly productive shelf sea, probably under slightly anoxic (low-oxygen) conditions, judging by the lack of bottom-dwelling organisms.
REMARK The genus *Orthoceras* itself is restricted to a single Ordovician species.

phragmocone

siphuncle

calcite-filled chambers

siphuncle

Paper weights/ Letter openers

Typical length 12 in (30 cm)

Orthocycloceras fluminense
(Meneghini); Late Silurian; Morocco.

Range: Silurian–Devonian	Distribution: Worldwide	Occurrence: ◉◉◉◉◉

| Group: NAUTILIDA | Subgroup: TRIGONOCERATIDAE | Informal name: Nautiloid |

Discitoceras

This nautiloid had an evolute shell with a widely open umbilicus and a rounded venter. A spiral ornament of fine, closely spaced ribs is present on the shell's exterior.

HABITAT *Discitoceras* lived around deep-water algal reefs in the Early Carboniferous seas and probably swam poorly, like the living pearly nautilus.

REMARK An external mold of another specimen is present on the outside of the living chamber.

mold of another fossil

fine, spiral ribs

rounded venter

Typical diameter
6 in (15 cm)

Discitoceras leveilleanum (de Koninck); Carboniferous Limestone; Early Carboniferous; Ireland.

| Range: Early Carboniferous | Distribution: N.W. Europe | Occurrence: ◉ |

| Group: PSEUDORTHOCERIDA | Subgroup: CARBACTINOCERATIDAE | Informal name: Nautiloid |

Rayonnoceras

The shell of *Rayonnoceras* is straight and conical, with a smooth surface. The centrally placed siphuncle has a complex structure. Both the siphuncle and the convex chambers are filled with calcareous deposits formed during the life of the nautiloid.

HABITAT *Rayonnoceras* was probably a bottom-dwelling ambush predator in a shallow-water marine environment.

REMARK The calcareous deposits in the shell are thought to have acted as ballast to make the animal heavier.

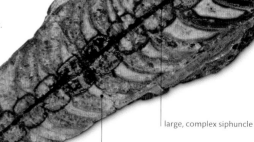

Rayonnoceras giganteum (J. Sowerby); Carboniferous Limestone; Early Carboniferous; Ireland.

large, complex siphuncle

chambers filled with calcareous deposits

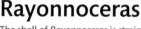

Typical diameter
8 in (20 cm)

| Range: Early Carboniferous | Distribution: Europe, N. America | Occurrence: ◉ ◉ |

| Group: NAUTILIDA | Subgroup: TRIGONOCERATIDAE | Informal name: Nautiloid |

Vestinautilus

The shell is notable for its extremely evolute form. It has an open umbilicus and a distinctive midflank ridge that runs along each side, on which the sutures form a forwardly directed V-shaped fold. There are numerous small chambers.

HABITAT *Vestinautilus* was a poor swimmer living on the bottom of shelf seas.

REMARK The specimen is preserved as an internal mold in limestone; the body chamber has broken away. The sutures are well displayed.

Vestinautilus caniferous (J. de C. Sowerby); Carboniferous; Locality unknown.

open umbilicus

folded suture

Typical diameter 2¹/₂ in (6 cm)

| Range: Early Carboniferous | Distribution: Europe, US, S. America | Occurrence: ◉◉ |

| Group: NAUTILIDA | Subgroup: NAUTILIDAE | Informal name: Nautilus |

Eutrephoceras

The coiling of the shell is very involute, the umbilicus extremely narrow. The outline of the shell is nearly globular; the venter is very broadly rounded. The individual chambers of the phragmocone are fewer than in the genus *Cenoceras* (see p.139).

HABITAT *Eutrephoceras* probably lived as a slow-moving predator in the rather shallow waters of the Western Interior Seaway, which occupied the Midwest of the US during much of the Late Cretaceous period.

REMARK This particular specimen is preserved as an internal mold in dark mudstone, on the surface of which patches of the original shell material adhere. These patches of shell have the "mother of pearl" colors that are characteristic of all cephalopod shells.

Eutrephoceras dekayi (Morton); Pierre Shale Formation; Late Cretaceous; US.

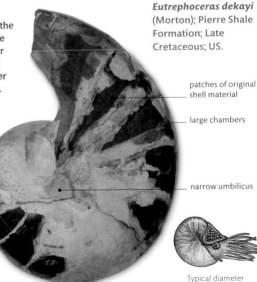

patches of original shell material

large chambers

narrow umbilicus

Typical diameter 4 in (10 cm)

| Range: Late Cretaceous | Distribution: US | Occurrence: ◉◉ |

| Group: NAUTILIDA | Subgroup: CENOCERATIDAE | Informal name: Nautilus |

Cenoceras

Internal mold

This is an extinct, long-ranging genus of nautiloid. The planispiral shell varies from loosely to tightly coiled and discoidal to globose in shape. The siphuncle is usually situated centrally in each compartment of the phragmocone. The outer surface is generally smooth but may be finely ribbed and ornamented in some species. The living chamber occupies about half a whorl.

HABITAT Like the living Nautilus, *Cenoceras* was likely to have been a marine predator and an opportunistic ocean-floor scavenger.

REMARK Nautiloids can change their buoyancy by altering the proportions of gas and liquid in their chambers. This allows them to vertically migrate from deep to shallow water at night. Many *Cenoceras* shells are broken, having imploded as a result of the shell sinking into deeper water after the animal died.

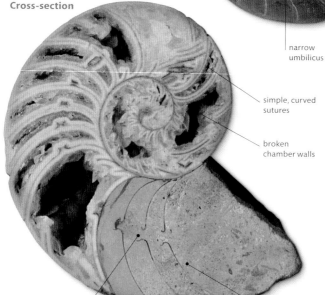

broad, rounded venter

mud-filled chambers

Cross-section

narrow umbilicus

simple, curved sutures

broken chamber walls

Cenoceras sp.; Inferior Oolite; Middle Jurassic; UK.

Cenoceras bradfordense (Crick); Inferior Oolite; Middle Jurassic; UK.

siphuncle joining chambers

start of living chamber

Typical diameter 6 in (15 cm)

| Range: Late Triassic–Middle Jurassic | Distribution: Worldwide | Occurrence: ● ● ● ● |

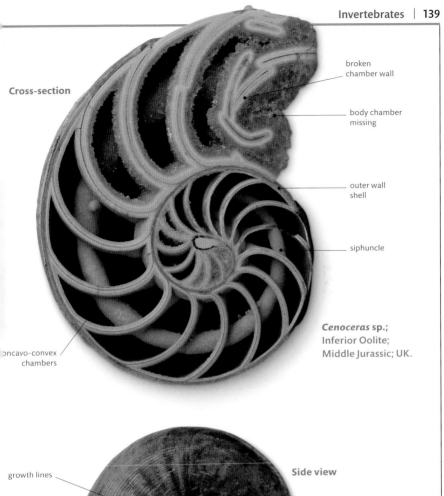

Cross-section

broken
chamber wall

body chamber
missing

outer wall
shell

siphuncle

Cenoceras sp.;
Inferior Oolite;
Middle Jurassic; UK.

oncavo-convex
chambers

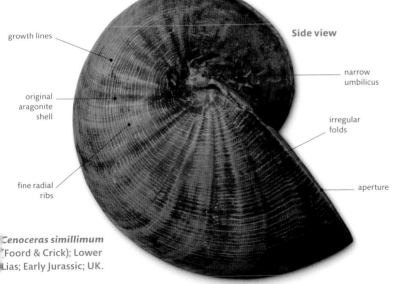

growth lines

original
aragonite
shell

fine radial
ribs

Side view

narrow
umbilicus

irregular
folds

aperture

Cenoceras simillimum
(Foord & Crick); Lower
Lias; Early Jurassic; UK.

Group: NAUTILIDA	Subgroup: ATURIIDAE	Informal name: Nautilus

Aturia

The shell is very involute and compressed, with a narrow, shallow umbilicus and a rounded venter. The suture line has a strong, backward-pointing fold on the flank, which gives it a superficial resemblance to some primitive Paleozoic ammonoids.

HABITAT *Aturia* probably lived in relatively deep water; the compressed shell was streamlined to allow fast movement.

REMARK The specimen is an internal mold of the chambered part of the shell.

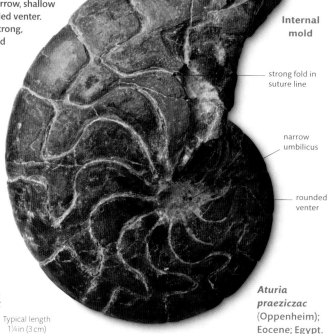

Internal mold

strong fold in suture line

narrow umbilicus

rounded venter

Aturia praeziczac (Oppenheim); Eocene; Egypt.

Typical length
1¼ in (3 cm)

Range: Paleocene–Miocene	Distribution: Worldwide	Occurrence: ◉

Group: NAUTILIDA	Subgroup: VARIOUS	Informal name: Rhyncholite

Rhyncholites

Rhyncholites is the genus given to the fossilized remains of the upper jaw of a nautiloid. Like the living nautilus, the nautiloids had two strong biting jaws, resembling the beak of a parrot. Composed of calcite, they preserved well. These jaws were ideal for cutting fish and crustacea.

Upper jaw

Rhyncholites sp.; Eocene; Libya.

Typical length
4¾ in (12 cm)

biting edge

Range: Triassic–Pliocene	Distribution: Worldwide	Occurrence: ◉◉

AMMONOIDS

AMMONOIDS EVOLVED from nautiloids in the early Devonian period, about 400 million years ago, and were abundant in world seas for the following 370 million years, after which they vanished suddenly at the end of the Cretaceous period. The rapid evolution of ammonoids and their widespread distribution makes them of great value in the subdivision of Late Paleozoic and Mesozoic time. As a group, they are characterized by the position of the siphuncle (the tube connecting the chambers of the shell), which is near the outside of the shell (ventral). Ammonoidea sutures can be simple, as found in Paleozoic species, or complex, as seen in Mesozoic species. Because they are extinct, we know very little about the soft parts and life habits of the ammonoids. It is rare to find the biting jaws; the tonguelike, rasping radula; or the ink sacs preserved in living chambers.

Group: CLYMENIIDA	Subgroup: CLYMENIIDAE	Informal name: Clymeniid

Clymenia

The evolute shell has a wide, open umbilicus. It is either nearly smooth or it carries weak, gently curved growth lines. The whorl section is compressed, and the venter is rounded. The suture is very simple.

HABITAT This genus lived a predatory existence on or near the floor of Devonian seas.
REMARK One of the more primitive ammonoids, *Clymenia* was locally common in Late Devonian rocks but not well preserved.

Clymenia laevigata (Münster); Late Devonian; Germany.

limestone matrix

broad umbilicus

rounded venter

Typical diameter 1½ in (4 cm)

Range: Late Devonian	Distribution: Europe, Asia, N. Africa	Occurrence: ◉◉

Group: CLYMENIIDA	Subgroup: GLATZIELLIDAE	Informal name: Clymeniid

Soliclymenia

This is an unusually shaped ammonoid, with a very evolute shell that shows distinctive triangular coiling. It has a broad umbilicus and a rounded venter. The simple ribs are closely spaced; the suture is very simple.

HABITAT It lived in moderately deep water. Although it could swim, its shell shape suggests it spent some time on the sea bed.

Soliclymenia paradoxa (Münster); Late Devonian; Germany.

simple, close ribs

evolute shell form

triangular coiling

Typical length ¾ in (2 cm)

Range: Late Devonian	Distribution: Eurasia, N. Africa, N. America	Occurrence: ◉◉

Group: PROLECANITIDA	Subgroup: PROLECANITIDAE	Informal name: Goniatite

Merocanites

The shell of *Merocanites* is evolute and has a broad, open umbilicus; the profile is compressed, the sides parallel. The gently rounded flanks show a suture with several blade-shaped folds (a fairly complex form for Paleozoic ammonoids). The venter is rounded.

HABITAT *Merocanites* lived in moderately deep water in the Early Carboniferous seas.
REMARK The specimen is preserved in dark limestone as an iron-oxide-coated internal mold. The living chamber is not preserved.

complex suture with deep folds

Internal mold

fine, dark, muddy limestone matrix

Merocanites compressus (J. Sowerby); Carboniferous Limestone; Early Carboniferous; UK.

Typical length 2 in (5 cm)

Range: Early Carboniferous	Distribution: Europe, Asia, N. America	Occurrence: ◉

| Group: CERATITIDA | Subgroup: CERATITIDAE | Informal name: Ceratite |

Ceratites

The shell is moderately evolute, with a broad umbilicus and a flat venter. The sides bear strong, simple, widely spaced ribs, which end on the ventral margin in a tubercle. The closely spaced suture lines have a highly distinctive form, with shallow, smoothly rounded folds facing anteriorly and toothed folds facing posteriorly.

HABITAT *Ceratites* lived in the shallow waters of the Triassic seas.

living chamber

limestone matrix

strong, simple ribs

suture

Internal mold

Ceratites nodosus (Bruguière); Triassic; Germany

Typical length 2½ in (6 cm)

| Range: Triassic | Distribution: Europe | Occurrence: ●●● |

| Group: GONIATITIDA | Subgroup: GONIATITIDAE | Informal name: Goniatite |

Goniatites

The shell is strongly involute, with a small, narrow umbilicus. The suture includes both pointed and rounded elements in the form of a zigzag. The thin shell has an ornament of fine, closely spaced growth lines.

HABITAT *Goniatites* lived in Carboniferous shelf seas, occurring locally in swarms over reef structures. The shape of its shell suggests that *Goniatites* was a poor swimmer.

Goniatites crenistria Phillips; Bowland Shales; Early Carboniferous; UK.

goniatitic suture line

broad, rounded venter

narrow, deep umbilicus

Typical length 2½ in (6 cm)

| Range: Early Carboniferous | Distribution: Worldwide | Occurrence: ●●● |

AMMONITES

AMMONITES ARE A FORM of ammonoid distinguished by their complex suture lines. They were abundant and diverse in the seas of the Mesozoic Era, and they evolved very rapidly to produce numerous species and genera. After a decline in diversity during the Late Cretaceous period, ammonites became extinct at the same time as other marine groups, such as belemnites, and terrestrial groups, such as dinosaurs. As both ammonites and all their close relatives are extinct, scientists know very little about their mode of life. What is known about them has been worked out mostly from experiments with model shells in water tanks.

Group: PHYLLOCERATIDA	Subgroup: PHYLLOCERATIDAE	Informal name: Phylloceratid

Phylloceras

A compressed, involute shell form and a distinctive frilly suture characterize the shells of these small- to medium-sized ammonites. They are either simply ornamented with growth lines or nearly smooth. The aperture is gently curved.

HABITAT The fairly streamlined profile of the shell and the rounded venter allowed this genus to swim at moderate speeds by jet propulsion.

Typical diameter 4 in (10 cm)

complex, frilled suture line

Polished internal cast

Phylloceras sp.; Early Jurassic; Mexico.

Range: Early Jurassic–Late Cretaceous	Distribution: Worldwide	Occurrence: ●●●●

Group: LYTOCERATIDA	Subgroup: LYTOCERATIDAE	Informal name: Lytoceratid

Lytoceras

The shell of *Lytoceras* is evolute, with a wide umbilicus. The whorl section is evenly rounded. The shell is ornamented with fine, closely spaced ribs and less frequent flanges, which pass uninterrupted over the venter. The suture is complex.

HABITAT The shell is ill-adapted for fast swimming, and it may have lived near the bottom of the sea.
REMARK Like *Phylloceras* (see above), *Lytoceras* is most abundant in low-latitude, deep-water deposits of the ancient Tethys Ocean.

Lytoceras fimbriatus (J. de C. Sowerby); Middle Lias; Early Jurassic; UK.

fine ribs

Typical length 4 in (10 cm)

Range: Early Jurassic	Distribution: Worldwide	Occurrence: ●●

| Group: AMMONITIDA | Subgroup: OXYNOTICERATIDAE | Informal name: Ammonite |

Oxynoticeras

living chamber
broken away

**Internal cast
in pyrites**

This distinctive ammonite is characterized by its strongly compressed, involute shell and knife-sharp keel. The suture line is complex and frilled. The lower specimen has been cut in half to show the septa (pyrite) and the cameral chambers, now filled with yellow calcite. The siphuncle is just visible next to the venter—its characteristic position.

HABITAT The sharp keel would have offered minimum resistance to the water through which *Oxynoticeras* swam, and it is interpreted as one of the fastest-swimming ammonites of all.

REMARKS The specimens are internal molds in bronzy iron pyrites—a common ammonite mode of preservation in clays.

Oxynoticeras oxynotum
(Quenstedt); Lower Lias;
Early Jurassic; UK.

involute shell with
small umbilicus

cross-section of
individual, nearly
straight septa

siphuncle just visible on
ventral margin of shell

chamber full of
yellow calcite

sharp keel
on margin of
compressed
shell

**Pyritized
ammonite cut
in half**

Typical diameter
4 in (10 cm)

| Range: Early Jurassic | Distribution: Worldwide | Occurrence: ⊛⊛⊛ |

Group: AMMONITIDA	Subgroup: PSILOCERATIDAE	Informal name: Ammonite

Psiloceras

In many localities around the world, this small, rather smooth genus is taken as the marker of earliest Jurassic time. The Psiloceratidae probably evolved from the *Phylloceras* group (see p.144) and retain rather simple sutures. The preservation of original deep pink, pearly shell, crushed flat in shale, is typical of fossil shells found in north Somerset, UK.

HABITAT *Psiloceras* was a moderately capable swimmer.
REMARKS This genus was abundant locally in the Early Jurassic period.

gray shale matrix

small individuals preserved as nacre shell

crushed shells

Typical diameter
2¼ in (7 cm)

Psiloceras planorbis
(J. de C. Sowerby); Lower Lias; Early Jurassic; UK.

Range: Early Jurassic	Distribution: Worldwide	Occurrence: ⓐⓐⓐⓐⓐ

Group: AMMONITIDA	Subgroup: AMALTHEIDAE	Informal name: Ammonite

Amaltheus

Amaltheus is characterized by its very compressed, involute shell and by the narrow keel, which has an ornament resembling a piece of rope. The ornament of the flanks is weak and consists of sickle-shaped ribs or, more rarely, fine, spiral ribs.

HABITAT The streamlined whorl profile and narrow keel suggest that it was quite a good swimmer.
REMARKS The genus may have evolved from *Phylloceras* (see p.144).

simple
S-shaped ribs

Brown
sandstone
mold

Typical diameter
2¼ in (7 cm)

Amaltheus stokesi
(J. Sowerby);
Middle Lias; Early Jurassic; UK.

compressed shell
with corded keel

Range: Early Jurassic	Distribution: Worldwide	Occurrence: ⓐⓐⓐⓐ

| Group: AMMONITIDA | Subgroup: DOUVILLEICERATIDAE | Informal name: Tractor Ammonite |

Douvilleiceras

The shell is involute and compressed in profile. Strong, simple ribs pass over the rounded venter and are each divided into numerous, even-sized tubercles, which would originally have carried short spines.

HABITAT This genus was certainly a poor swimmer—the broad profile of the whorl offered considerable resistance to the water. *Douvilleiceras* may have spent a lot of time scavenging or hunting on the sea bed.
REMARK This well-preserved specimen from Madagascar has retained most of its original nacreous shell.

Douvilleiceras inaequinodum (Quenstedt); Early Cretaceous; Madagascar.

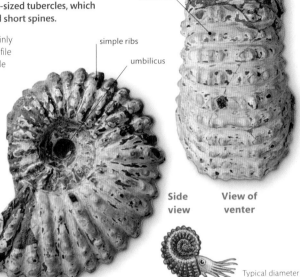

tubercles

simple ribs

umbilicus

Side view

View of venter

Typical diameter 4 in (10 cm)

| Range: Early Cretaceous | Distribution: Worldwide | Occurrence: ⊛⊛⊛⊛ |

| Group: AMMONITIDA | Subgroup: PERISPHINCTIDAE | Informal name: Ammonite |

Perisphinctes

The shell is evolute, exposing many whorls and a wide, shallow umbilicus. Finely spaced branching ribs pass over the rounded venter without interruption; there is no keel present.

HABITAT *Perisphinctes* lived in warm, shallow shelf seas and was probably a relatively slow swimmer.
REMARKS This species and *Douvilleiceras* (above) are commercially mined for the fossil trade.

Typical diameter 4 in (10 cm)

Perisphinctes (*Dichotomoceras*) *virguloides* (Waagen); Late Jurassic; Madagascar.

bifid ribs

aperture

venter

Side view

umbilicus

View of venter

| Range: Middle–Late Jurassic | Distribution: Worldwide | Occurrence: ⊛⊛⊛ |

| Group: AMMONITIDA | Subgroup: KOSMOCERATIDAE | Informal name: Ammonite |

Kosmoceras

This is a compressed, moderately evolute ammonite. It is very typical of Middle Jurassic deposits. The shell has a complex ornament of bunched ribs and rows of tubercles. The venter is narrow and flat.

HABITAT *Kosmoceras* lived in deeper water and was probably a moderately good swimmer.
REMARK The specimen is preserved in original shell material, crushed in shales.

Kosmoceras duncani
(J. de C. Sowerby); Oxford Clay; Middle Jurassic; UK.

crushed nacre shell

Typical diameter
2¹/₂ in (6 cm)

| Range: Middle Jurassic | Distribution: Worldwide | Occurrence: ⓐⓐⓐⓐⓐ |

| Group: AMMONITIDA | Subgroup: PERISPHINCTIDAE | Informal name: Ammonite |

Pavlovia

This is a typical member of the abundant and widespread Late Jurassic family, Perisphinctidae. Members of the family are characterized by open, evolute shell forms; rounded whorl sections; and ornamentation of branching ribs. A long living chamber occupies nearly a whole whorl of the shell. The suture is complex.

HABITAT This genus lived in the extensive Late Jurassic seas.

branching ribs

evolute shell form

umbilicus

Typical diameter
1¹/₂ in (4 cm)

Pavlovia sp.;
Upper Glauconite Series;
Late Jurassic;
Greenland.

| Range: Late Jurassic | Distribution: Greenland, northern Europe | Occurrence: ⓐⓐⓐ |

| Group: AMMONITIDA | Subgroup: DACTYLIOCERATIDAE | Informal name: Ammonite |

Dactylioceras

The shell form is evolute, with many whorls visible and a wide, shallow umbilicus. The fine, branching ribs are closely spaced and pass over a rounded venter.

HABITAT Experiments have shown *Dactylioceras* to be a slow swimmer.

REMARK In medieval times, ammonites were thought to be petrified snakes and were provided with carved heads for sale to pilgrims.

aperture

carved "head"

Dactylioceras commune (J. de C. Sowerby); Upper Lias; Early Jurassic; UK.

open-coiled, evolute form

Typical diameter 2³/₄ in (7 cm)

fine, branching ribs

rounded venter

| Range: Early Jurassic | Distribution: Worldwide | Occurrence: ⦾⦾⦾⦾⦾ |

| Group: AMMONITIDA | Subgroup: HILDOCERATIDAE | Informal name: Ammonite |

Harpoceras

This is a common Early Jurassic ammonite, characterized by a compressed, moderately involute shell, which bore a sharp ventral keel, and flanks with distinctive sickle-shaped ribs. The aperture was drawn into a short rostrum on the ventral margin.

HABITAT This genus was probably a good swimmer and fed by preying on small animals.

REMARK The specimen is preserved as claystone coated with a thin layer of iron pyrites.

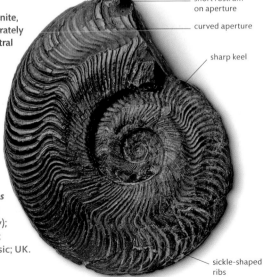

short rostrum on aperture

curved aperture

sharp keel

Harpoceras falciferum (J. Sowerby); Upper Lias; Early Jurassic; UK.

Typical diameter 4³/₄ in (12 cm)

sickle-shaped ribs

| Range: Early Jurassic | Distribution: Worldwide | Occurrence: ⦾⦾⦾⦾ |

| Group: AMMONITIDA | Subgroup: HILDOCERATIDAE | Informal name: Ammonite |

Hildoceras

Hildoceras had an evolute, laterally compressed shell. The coarse, widely spaced, sickle-shaped ribbing on the sides is interrupted by a groove. The suture line is very complex. It has a very distinctive rectangular whorl section, which on the venter carries three low keels separated by two grooves.

HABITAT *Hildoceras* was a moderate swimmer in the Early Jurassic shelf seas.
REMARK This specimen is well preserved in a claystone nodule.

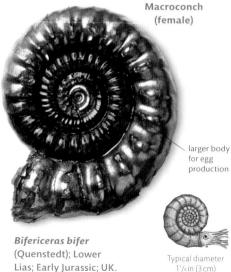

sickle-shaped ribs interrupted by groove

wide umbilicus

triple-keeled venter

Typical diameter
2³/₄ in (7 cm)

Hildoceras bifrons
(Bruguière); Upper Lias;
Early Jurassic; UK.

| Range: Early Jurassic | Distribution: Europe, Anatolia, Japan | Occurrence: ⊛ ⊛ ⊛ |

| Group: AMMONITIDA | Subgroup: EODEROCERATIDAE | Informal name: Ammonite |

Bifericeras

Macroconch (female)

In common with many ammonites, the larger shell (macroconch) of *Bifericeras* is the female, and the smaller (microconch) is the male. Females required a larger body size for egg production and possible brooding habit.

HABITAT *Bifericeras* lived in open seas of moderate depth and either predated on or scavenged small marine invertebrates.

larger body for egg production

Microconch (male)

Bifericeras bifer
(Quenstedt); Lower
Lias; Early Jurassic; UK.

Typical diameter
1¹/₄ in (3 cm)

| Range: Early Jurassic | Distribution: Europe | Occurrence: ⊛ ⊛ |

Group: AMMONITIDA	Subgroup: PACHYDISCIDAE	Informal name: Ammonite

Pachydiscus

The family Pachydiscidae includes the giants of the ammonite group, and specimens up to 6½ ft (2 m) in diameter have been found in Late Cretaceous rocks. The shell is moderately involute and compressed and carries short, gently curved ribs. The venter is rounded.

HABITAT A competent swimmer, *Pachydiscus* lived in open seas.
REMARK The specimen is preserved in limestone but retains patches of the original shell.

limestone-infilled living chamber

involute shell form

rounded venter

original shell material

Pachydiscus sp.; Late Cretaceous; Canada.

Typical diameter 2½ in (6 cm)

Range: Late Cretaceous	Distribution: Worldwide	Occurrence: ◗ ◗

Group: AMMONITIDA	Subgroup: CRIOCERATITIDAE	Informal name: Ammonite

Crioceratites

The shell of *Crioceratites* is so loosely coiled that the whorls are not in contact. The whorl size increases rapidly. Stronger ribs carrying tubercles are separated by two to three weaker, finer ribs with no tubercles. The venter is rounded.

HABITAT This genus was adapted to a predatory swimming habit in moderately deep, shelf seas.
REMARK The loosely coiled morphology evolved many times in the ammonoids and nautiloids.

open-coiled whorls

ribs with tubercles

fine ribs

Crioceratites sp.; Early Cretaceous; South Africa.

Typical diameter 4 in (10 cm)

Range: Early Cretaceous	Distribution: Worldwide	Occurrence: ◗ ◗ ◗

| Group: AMMONITIDA | Subgroup: ACANTHOCERATITIDAE | Informal name: Ammonite |

Mammites

The shell is robust, with a squarish whorl section and a flattened venter. The suture is relatively simple. There are three rows of rounded tubercles on stout ribs.

HABITAT *Mammites* inhabited the shallow, warm shelf seas of the Late Cretaceous.
REMARK They are mined commercially in Morocco. They occur as oyster-encrusted phosphatic internal casts, often in nodules.

strong, rounded tubercles

phosphatic internal cast

Mammites nodosoides (Schluter); Late Cretaceous; Morocco.

coarse ribbing

Typical diameter 6 in (15 cm)

| Range: Late Cretaceous | Distribution: Worldwide | Occurrence: ●●●● |

| Group: AMMONITIDA | Subgroup: DESMOCERATITIDAE | Informal name: Ammonite |

Aioloceras

The shell is discoidal, with a smooth, arched venter. The outer whorl covers approximately half the preceding one. The ribbing is rounded and indistinct and confined to the inner whorls. The sutures are frilled and very detailed.

HABITAT *Aioloceras* inhabited the shallow, warm shelf seas of the Early Cretaceous.
REMARK This species is mined commercially in Madagascar, where they are often sliced and polished for export.

hollow, calcite-lined chamber

original nacreous shell

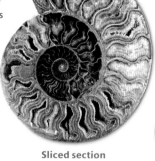

Sliced section

Polished internal cast

Aioloceras besairiei (Collignon); Early Cretaceous; Madagascar.

Typical diameter 8 in (20 cm)

complex, frilled sutures

| Range: Early Cretaceous | Distribution: Worldwide | Occurrence: ●●● |

| Group: AMMONITIDA | Subgroup: PLACENTICERATIDAE | Informal name: Ammonite |

Placenticeras

This Late Cretaceous genus grew to a considerable size (up to 20 in/50 cm)—the specimen shown here lacks the living chamber of the ammonite, which in life would have comprised almost another whorl. The highly compressed, very involute shell has a very narrow venter and a thick, nacreous shell.

HABITAT Experiments in tanks with models of ammonite shells indicate that *Placenticeras*, with its large size and very streamlined profile, was possibly the fastest swimmer of all known ammonites, jetting through the water in pursuit of prey or fleeing from reptiles and large fish.

REMARK Several specimens recovered from the Western Interior Seaway of the US display bite marks from the repeated attacks of a giant marine reptile.

tabulate venter

compressed profile

living chamber broken away

broken end of convoluted septum

Whorl profile

Placenticeras meeki (Böhm); Late Cretaceous; US.

former position of aperture

umbilicus

smooth, almost flat flank

faint ribs on flank

Typical diameter 8 in (20 cm)

| Range: Late Cretaceous | Distribution: Worldwide | Occurrence: ◉◉◉ |

Group: AMMONITIDA	Subgroup: EODEROCERATIDAE	Informal name: Ammonite

Promicroceras

This limestone (known as Marston Marble) is composed largely of closely packed, small but complete shells of the ammonite genus *Promicroceras*. A fragment of a larger genus, *Asteroceras*, is present on the corner of the block. *Promicroceras's* shell is evolute and is ornamented by coarse, simple, straight ribs, which form a forward-directed flange over the venter.

HABITAT Marston Marble was formed by the mass mortality of *Promicroceras*, resulting either from a major storm or possibly poisoning by an algal bloom.
REMARK The ammonite shells were rapidly buried in lime mud, which hardened into limestone.

Asteroceras shell

small, perfectly preserved shells of *Promicroceras*

original shell

Marston marble; Lower Lias; Early Jurassic; UK.

Typical diameter ³⁄₄ in (2 cm)

Range: Early Jurassic	Distribution: Worldwide	Occurrence: ◉◉

Group: AMMONITIDA	Subgroup: AREITITIDAE	Informal name: Ammonite

Asteroceras

Large, rather simply ornamented ammonites of this family are common in Early Jurassic sediments on a worldwide scale. The rather evolute shell carries curved ribs, and a keel is present on the venter.

HABITAT *Asteroceras* lived at moderate depths in shelf seas and was probably a slow swimmer, living by predation on small marine animals.
REMARK The chambered part of this well-preserved shell has infilled with translucent brown calcite, and the living chamber with gray lime mud. The shell has mostly broken away.

Asteroceras obtusum (J. Sowerby); Lower Lias; Early Jurassic; UK.

simple ammonite sutures

chamber that housed the soft parts

Typical diameter 8¹⁄₂ in (22 cm)

Range: Early Jurassic	Distribution: Worldwide	Occurrence: ◉◉◉

Group: AMMONITIDA	Subgroup: ECHIOCERATIDAE	Informal name: Ammonite

Echioceras

The shell is extremely evolute, with numerous visible whorls and a shallow umbilicus. Short, straight ribs are present on the flanks but do not pass on to the low-keeled venter.

HABITAT This ammonite inhabited the shelf seas of the Early Jurassic. From the very evolute shell shape, it is possible to infer that *Echioceras* was not adapted for fast swimming. It probably scavenged or caught slow-moving prey.

venter carries low keel

short, straight ribs on evolute shell

Typical diameter 2½ in (6 cm)

Echioceras **sp.; Lower Lias; Early Jurassic; UK.**

Range: Early Jurassic	Distribution: Worldwide	Occurrence: ◉◉◉◉

Group: AMMONITIDA	Subgroup: AMALTHEIDAE	Informal name: Ammonite

Pleuroceras

The shell is planate, with about a 25 percent overlap, exposing many whorls and a relatively open umbilicus. The whorl section is quadrate, with strong, evenly situated radial ribs and a prominent corded keel on a flattened venter.

HABITAT This was a small, fast-moving nektonic ammonite, only known from the shallow shelf seas of the northern hemisphere.
REMARK This specimen is preserved in solid iron pyrite but retains a thin layer of its original nacreous shell.

rib

corded keel   venter

Typical diameter 2 in (5 cm)

Pleuroceras spinatum (Bruguière); **Early Jurassic; Germany.**

Side view

umbilicus

View of venter

Range: Early Jurassic	Distribution: Europe, North America	Occurrence: ◉◉◉◉

| Group: AMMONITIDA | Subgroup: ACANTHOCERATIDAE | Informal name: Ammonite |

Mantelliceras

The shell is involute, with a fairly small umbilicus. The whorl section is rectangular, the venter flat. Alternately long and short ribs are present on the flanks.

HABITAT *Mantelliceras* and its relatives were probably poor swimmers, because they were not streamlined. They display strong sexual dimorphism in size, with the females (macroconchs) much larger.

REMARK The genus is characteristic of the beginning of the Late Cretaceous in the northern hemisphere.

strong, intercalated, long and short ribs

simple aperture of large female shell

broad venter

Mantelliceras sp.; Lower Chalk; Late Cretaceous; UK.

Typical diameter
1³/₄ in (4.5 cm)

| Range: Cretaceous | Distribution: Europe | Occurrence: ◉◉ |

| Group: AMMONITIDA | Subgroup: HOPLITIDAE | Informal name: Ammonite |

Euhoplites

The shell is moderately involute and inflated in profile. The flanks bear tubercles, from which arise bunched ribs. The venter has a narrow groove.

HABITAT The numerous ribs and tubercles, together with the rectangular profile of the whorl, would have offered much resistance to water and prevented *Euhoplites* from swimming fast.

REMARK In life, the shell would have been three times the size of this fragment, which has missing whorls.

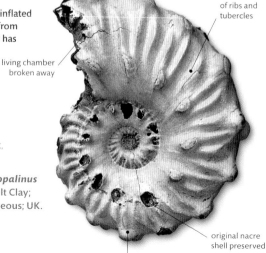

strong ornament of ribs and tubercles

living chamber broken away

Euhoplites opalinus (Spath); Gault Clay; Early Cretaceous; UK.

Typical diameter
1³/₈ in (3.5 cm)

original nacre shell preserved

pyrite preservation

| Range: Early Cretaceous | Distribution: Europe | Occurrence: ◉◉◉◉ |

Group: AMMONITIDA	Subgroup: SCHLOENBACHIIDAE	Informal name: Ammonite

Schloenbachia

Schloenbachia is one of the most typical Late Cretaceous ammonites of Europe. It displays considerable variation in the development of ornament—some specimens are nearly smooth and very flat; others are fat, with tubercles. The strongly developed ventral keel is characteristic of the genus. Individuals can grow up to 10 in (25 cm) across.

HABITAT Ammonites of this genus are thought to be fairly good swimmers.
REMARK This specimen is preserved as a phosphatic internal cast.

ventral keel

strong, rounded tubercles

Whorl profile

rows of well-developed tubercles on flanks

broken end of septum

Typical diameter 2 in (5 cm)

Schloenbachia varians (J. de C. Sowerby); Lower Chalk; Late Cretaceous; UK.

Range: Late Cretaceous	Distribution: Europe, Greenland	Occurrence:

Group: AMMONITIDA	Subgroup: BRANCOCERATIDAE	Informal name: Ammonite

Mortoniceras

The genus *Mortoniceras* is a very widespread form in Early Cretaceous sediments and may grow to 20 in (50 cm). The highly characteristic shell form includes a strong keel on the venter and a squarish whorl section.

HABITAT The genus was probably a poor swimmer and would have swum slowly through the water.
REMARK The specimen has been crushed into an elliptical shape in the rock.

Mortoniceras potternense (Spath); Early Cretaceous; UK.

irregular tubercles set on ribs

evolute, open-coiling style

strong ventral keel

Typical diameter 3¹/₄ in (8 cm)

Range: Early Cretaceous	Distribution: Europe, Africa, US	Occurrence:

Group: AMMONITIDA	Subgroup: DESHAYESITIDAE	Informal name: Ammonite

Deshayesites

Deshayesites was an evolute, compressed ammonite whose shell carries a rather simple ornament of sinuous ribs. The ribs pass over the narrow, rounded venter. The genus was common in, and absolutely diagnostic of, the Aptian stage.

HABITAT The streamlined profile and weak ornament suggest that this genus was capable of fast movement through the water.
REMARK The specimen is preserved as an internal mold with traces of original shell.

sinuous ribs

broken end of living chamber

rounded edge of venter

Typical diameter
1¾ in (4.5 cm)

Deshayesites forbesi Casey; Early Cretaceous; Russia.

Range: Early Cretaceous	Distribution: Worldwide	Occurrence: ◉◉◉

Group: AMMONITIDA	Subgroup: NOSTOCERATIDAE	Informal name: Ammonite

Nipponites

Of all the Late Cretaceous ammonites that coiled in an irregular fashion, *Nipponites* was the most bizarre. At first sight, the shell appears to be an irregular tangle of whorls. On closer inspection, however, it proves to be a 3D network of Us. It possesses a typical complex ammonite suture and a simple ornament of ribs, and it probably evolved from a helically coiled form (see p.159).

HABITAT *Nipponites* probably lived as a planktonic form, drifting through the midlevel or upper waters of the warm Late Cretaceous seas and feeding on small animals, which it caught with its tentacles.

Brown sandstone internal cast

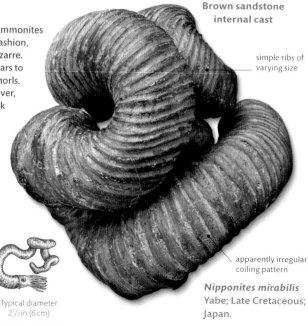

simple ribs of varying size

apparently irregular coiling pattern

Typical diameter
2½ in (6 cm)

Nipponites mirabilis Yabe; Late Cretaceous; Japan.

Range: Late Cretaceous	Distribution: Japan, US	Occurrence: ◉

| Group: AMMONITIDA | Subgroup: NOSTOCERATIDAE | Informal name: Ammonite |

Bostrychoceras

In this genus, and in the closely related family Turrilitidae, the coiling of the shell has become helical (like a snail or gastropod), in contrast to the planispiral, flat coiling of most ammonites. Although superficially similar to gastropods, such fossils are readily identifiable as ammonites from the characteristic suture lines. In *Bostrychoceras*, the coiling is loose so that successive whorls are not in contact. The simple ornament of fine, closely spaced ribs is interrupted on the living chamber by the development of tubercles. The living chamber is U-shaped, and the aperture was directed forward in life so that the protruding tentacles would not have been in contact with the sea bed below.

HABITAT Paleontologists believe that this genus was probably planktonic, floating in the open ocean and feeding on small animals in the water column. The very widespread distribution matches up quite well with this theory, because ocean-going species are commonly of global occurrence. Another possibility is that they were bottom dwelling and moved along the seafloor hunting for small prey rather like an octopus.
REMARK *Bostrychoceras* is most often found as small pieces of broken whorls, the ornamentation of which allows identification.

Yellow limestone cast

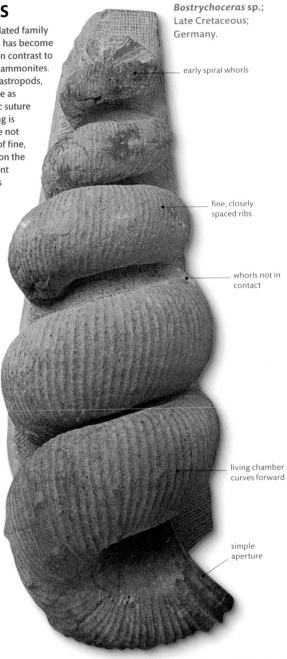

Bostrychoceras sp.; Late Cretaceous; Germany.

early spiral whorls

fine, closely spaced ribs

whorls not in contact

living chamber curves forward

simple aperture

Typical diameter
5¹/₂ in (14 cm)

| Range: Late Cretaceous | Distribution: Worldwide | Occurrence: 🌑🌑 |

Group: AMMONITIDA	Subgroup: BACULITIDAE	Informal name: Ammonite

Baculites

In this Late Cretaceous genus, only the very earliest part of the shell remained coiled; the latter part grew into a straight shaft. *Baculites* can occur in vast numbers at some localities, often to the virtual exclusion of other species, and could grow to over 39 in (1 m) in length.

HABITAT The mode of life of these straight ammonites is controversial: some paleontologists believe they lived upright in the water, with the tentacles on the sea bed foraging for food, while others think they had horizontal orientation and lived closer to the surface of the sea.

fragment of chambered shell

complex suture

traces of suture

living chamber

Typical length 4 in (10 cm)

Baculites sp.;
Calcaire a Baculites;
Late Cretaceous; France.

Range: Late Cretaceous	Distribution: Worldwide	Occurrence: ◉◉◉◉

Group: AMMONITIDA	Subgroup: SCAPHITIDAE	Informal name: Ammonite

Jeletzkytes

The family Scaphitidae represents a different design of Late Cretaceous uncoiled ammonite, in which the chambered early whorls are tightly coiled in the usual fashion but the living chamber, comprising a short, straight shaft and a terminal hook, is shaped like a shepherd's crook.

HABITAT The shell form of *Jeletzkytes* suggests it was unable to swim actively but could use gas and liquid in its chambered shell to alter its buoyancy and thus the position in the water column, in common with other ammonite genera.

REMARK This specimen was preserved in a mudstone concretion full of small, bivalve mollusks.

fine ribs   tubercles

original nacreous shell

Typical diameter 4 in (10 cm)

Jeletzkytes nebraskensis
(Owen); Late Cretaceous; US.

body chamber

Range: Late Cretaceous	Distribution: Worldwide	Occurrence: ◉◉◉

BELEMNITES AND SQUIDS

BELEMNITES AND SQUIDS belong to a diverse group of cephalopods, which include the living squid, cuttlefish, and octopus. The group is characterized by an internal, chambered shell enclosed entirely by soft, muscular tissues. Many forms have a hard internal support structure made of calcium carbonate or protein, known as the pen or guard. Belemnites possessed strong, usually cylindrical calcite guards that preserved well and occur abundantly in Mesozoic marine rocks. Squids possess an internal support called a pen, which is made of proteinaceous chitin. This is only rarely preserved. However, in a few instances, the soft tissue has been preserved and an entire squid, complete with tentacles, is recognizable.

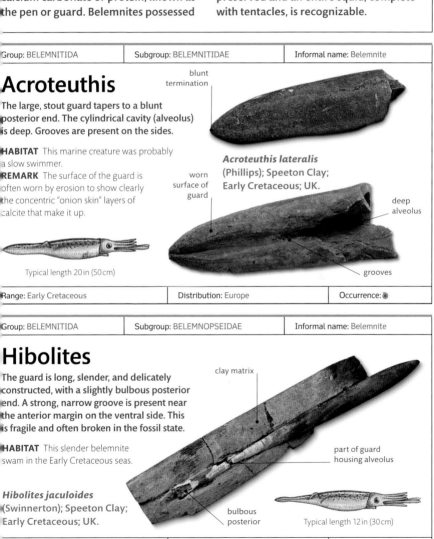

Group: BELEMNITIDA	Subgroup: BELEMNITIDAE	Informal name: Belemnite

Acroteuthis

blunt termination

The large, stout guard tapers to a blunt posterior end. The cylindrical cavity (alveolus) is deep. Grooves are present on the sides.

HABITAT This marine creature was probably a slow swimmer.
REMARK The surface of the guard is often worn by erosion to show clearly the concentric "onion skin" layers of calcite that make it up.

Acroteuthis lateralis (Phillips); Speeton Clay; Early Cretaceous; UK.

worn surface of guard

deep alveolus

Typical length 20 in (50 cm)

grooves

Range: Early Cretaceous	Distribution: Europe	Occurrence: ◉

Group: BELEMNITIDA	Subgroup: BELEMNOPSEIDAE	Informal name: Belemnite

Hibolites

The guard is long, slender, and delicately constructed, with a slightly bulbous posterior end. A strong, narrow groove is present near the anterior margin on the ventral side. This is fragile and often broken in the fossil state.

clay matrix

HABITAT This slender belemnite swam in the Early Cretaceous seas.

part of guard housing alveolus

Hibolites jaculoides (Swinnerton); Speeton Clay; Early Cretaceous; UK.

bulbous posterior

Typical length 12 in (30 cm)

Range: Middle Jurassic–Late Cretaceous	Distribution: Northern hemisphere	Occurrence: ◉

| Group: BELEMNITIDA | Subgroup: BELEMNOPSEIDAE | Informal name: Belemnite |

Neohibolites

The guard of this small belemnite is slender and spindle shaped, preserved as translucent, amber-colored calcite. It has a deep, rounded alveolus and an anterior groove. The proostracum is a tongue-shaped extension at the front of the phragmocone.

HABITAT *Neohibolites* lived in vast numbers, hunting small prey in the warm shelf seas of the mid-Cretaceous.

REMARK The separate chambers of the phragmocone on this specimen are still visible. *Neohibolites* was a very common fossil in the Albian clays of Europe.

chambered phragmocone

phragmocone sits in the alveolus of the guard

Neohibolites minimus (Miller); Gault Clay; Early Cretaceous; UK.

Typical length 9 in (23 cm)

| Range: Cretaceous | Distribution: Worldwide | Occurrence: ◉◉◉◉◉◉ |

| Group: BELEMNITIDA | Subgroup: CYLINDROTEUTHIDIDAE | Informal name: Belemnite |

Pachyteuthis

The guard of this large, stout belemnite tapers evenly to a blunt posterior termination. The broken surface of the guard displays its fibrous calcite construction. The rounded alveolus is deep and houses a large phragmocone, which consists of concavo-convex chambers now full of hardened mud.

HABITAT *Pachyteuthis* lived as a predator or scavenger in the deeper-water shelf seas of the Late Jurassic.

REMARK The phragmocone is well preserved in shiny iron pyrites.

blunt tip

worn surface of guard

chambered phragmocone preserved in iron pyrites

Pachyteuthis abbreviata (Miller); Kimmeridge Clay; Late Jurassic; UK.

V-shaped alveolus

Typical length 20 in (50 cm)

| Range: Middle–Late Jurassic | Distribution: Worldwide | Occurrence: ◉◉◉ |

Group: BELEMNITIDA	Subgroup: BELEMNITELLIDAE	Informal name: Belemnite

Belemnitella

The guard tapers only gently and is terminated posteriorly by a tip called a mucron. The anterior part of the guard is paper-thin and houses a deep alveolus. The surface of the guard carries a network of intricately branching, shallow grooves, which are probably impressions created by the blood vessels in the soft tissues of the living belemnite.

Belemnitella mucronata (Schlotheim); Late Cretaceous; The Netherlands.

HABITAT *Belemnitella* swam in the shallower waters of the Late Cretaceous Chalk sea and used its hooked tentacles to catch small prey.

REMARK This is one of the belemnites which survived nearly to the end of the Cretaceous. It occurs in huge numbers at some localities in so-called "belemnite graveyards."

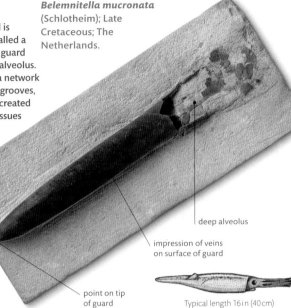

deep alveolus

impression of veins on surface of guard

point on tip of guard

Typical length 16 in (40 cm)

Range: Late Cretaceous	Distribution: Northern hemisphere	Occurrence: ◉◉◉◉◉

Group: BELEMNITIDA	Subgroup: CYLINDROTEUTHIDIDAE	Informal name: Belemnite

Cylindroteuthis

The guard is long and cylindrical and tapers gradually to a posterior point. The chambered phragmocone expands anteriorly.

Cylindroteuthis puzosiana (d'Orbigny); Oxford Clay; Late Jurassic; UK.

apex of guard

HABITAT This genus lived as a predator in the deeper parts of shelf seas.
REMARK This is one of the largest species of belemnite, growing up to 10 in (25 cm) in length.

long, cylindrical guard

alveolus of guard

Typical length 10 in (25 cm)

Range: Middle–Late Jurassic	Distribution: Europe, N. America	Occurrence: ◉◉◉◉

Group: Not applicable	Subgroup: Not applicable	Informal name: Belemnite limestone

Belemnite limestone

This cut and polished piece of marine shelly limestone shows a belemnite in cross-section. It illustrates well the massive construction of the belemnite guard, composed of the mineral calcite (calcium carbonate). Belemnites show fine, radially arranged, calcite crystals and display strong, concentric growth lines. The V-shaped space at the anterior end of the guard is called the alveolus, and it housed the chambered phragmocone, used for buoyancy control by the belemnite in life.

HABITAT Various belemnites would have had differing lifestyles, although all would have been marine dwellers.
REMARK The term "marble" is loosely applied to any polished limestone.

Treuchtlinger marble; Late Jurassic; Germany.

concentric growth rings

alveolus

calcite guard of *Hibolites* sp.

shell fragments in limestone

phragmocone

Range: Triassic–Cretaceous	Distribution: Worldwide	Occurrence: ◉◉◉◉◉

Group: TEUTHIDA	Subgroup: TRACHYTEUTHIDIDAE	Informal name: Belemnite

Trachyteuthis

Only the guard of this genus has ever been found preserved as a fossil. The outline is an elongated oval with a lobe present on each side. On the dorsal surface, growth lines are visible.

HABITAT This animal probably resembled a cuttlefish in its life habits, living on the sea bed and preying on crustaceans.
REMARK It is open to debate whether this genus is closer to the squid or cuttlefish in affinity.

***Trachyteuthis* sp.; Solnhofen Limestone; Late Jurassic; Germany.**

Typical length 10 in (25 cm)

growth lines on surface

flattened guard resembling cuttlebone

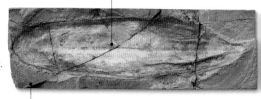

limestone

Range: Late Jurassic	Distribution: Europe	Occurrence: ◉

| Group: BELEMNITIDA | Subgroup: BELEMNOTEUTHIDAE | Informal name: Extinct squid |

Acanthoteuthis

The head region carries 10 arms, on which hooks are present. The anterior part of the body is called the mantle and has two fins—very much like a living squid. The hindmost part of the animal is the chambered phragmocone, which tapers posteriorly. At the end, a thin guard is sometimes preserved.

HABITAT *Acanthoteuthis* lived in the open sea at a moderate depth and swam by jet propulsion (water expulsion from the mantle) in exactly the same way as squid do in the present day. It used its hooked and suckered arms to catch prey.

REMARK This specimen displays a very rare phenomenon in the fossil record—the preservation of soft parts, which normally would decay and be completely lost. This type of preservation is caused by rapid burial of the animal in stagnant, oxygen-free water, where scavengers and most bacteria cannot live. The belemnite's soft parts are replaced by calcium phosphate.

arms with small hooks

head region

Acanthoteuthis antiqua (Pearce); Oxford Clay; Late Jurassic; UK.

Typical length 5 in (12 cm)

chambered phragmocone

mantle with two fins

| Range: Late Jurassic | Distribution: Europe | Occurrence: ◉ |

CRINOIDS

CRINOIDS, POPULARLY known as sea lilies and feather stars, possess a massive calcite skeleton and were so abundant in the Paleozoic seas that their remains formed vast thicknesses of limestone. Most crinoids are attached to the sea bed by a flexible stem, circular or pentagonal in section, and made up of numerous disklike plates called columnals. At the top of the stem is a swollen cup or calyx, to which the arms are attached. The arms are used to filter food from the water. Soon after death, the entire skeleton normally falls apart into small, separate plates called ossicles. In contrast, well-preserved crinoids are rare and beautiful fossils.

Group: MONOBATHRIDA	Subgroup: SCYPHOCRINITIDAE	Informal name: Sea lily

Scyphocrinites

This large crinoid has a 10-ft (3-m) long stalk with an ornamented cup and arms extending a further 12 in (30 cm). The stem is circular in cross-section, reducing in diameter away from the crown and terminating in a large, globular lobolith. This is hollow, divided into compartments and composed of tangled cirri (fine tendrils) in some species or small interlocking plates in others.

HABITAT This would explain their wide geographic distribution. However, recently, this idea has been challenged by the suggestion that they lived on the sea floor and the lobolith acted as a drag anchor.
REMARK *Scyphocrinites* itself is restricted to Silurian rocks. This lobolith (right) is a plate lobolith from the early Devonian and belongs to a close relative of *Scyphocrinites*.

feathery arms

Two cups

calyx or cup

stalk

wall of small plates

stem attachment

Scyphocrinites elegans (Zekner); Late Silurian; Morocco.

Lobolith

Typical cup diameter 3 in (7.5 cm)

Range: Silurian	Distribution: Europe, N. America	Occurrence: ◉◉◉

Group: CLADIDA	Subgroup: CUPRESSOCRINITIDAE	Informal name: Sea lily

Cupressocrinites

This robust crinoid has a cylindrical stem and a large, conical cup made up of 10 large plates in two alternating circles. The short arms, five in number, are unbranched and triangular in shape. They fit together closely to form a short, pointed crown.

HABITAT The typically robust form of this genus is possibly related to the rough, current-swept environment in which it lived, where more delicate forms would not have survived.

REMARK The ancestry and relationships of *Cupressocrinites* are not well understood.

compact crown

cup made of 10 large plates

cylindrical stem

Cupressocrinites crassus (Goldfuss); Crinoiden Schichten; Middle Devonian; Germany.

Typical cup diameter 1 in (2.5 cm)

Range: Devonian	Distribution: W. Europe	Occurrence: ◉

Group: CLADIDA	Subgroup: CYATHOCRINITIDAE	Informal name: Sea lily

Cyathocrinites

Cyathocrinites has a cylindrical stem made up of numerous disk-shaped columnals. The bowl-shaped cup includes a few smoothly rounded plates, and the narrow arms branch regularly to form a large, complex crown composed of many small units.

HABITAT This genus lived in calm, shallow water, which provided an abundant supply of food.

symmetrically branching arms

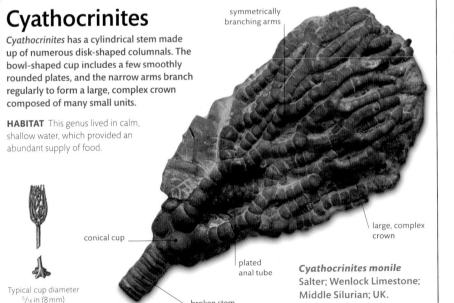

conical cup

large, complex crown

plated anal tube

Typical cup diameter 5/16 in (8 mm)

broken stem

Cyathocrinites monile Salter; Wenlock Limestone; Middle Silurian; UK.

Range: Silurian–Carboniferous	Distribution: Europe	Occurrence: ◉ ◉

Group: SAGENOCRINIDA	Subgroup: SAGENOCRINIDAE	Informal name: Sea lily

Sagenocrinites

The stem of *Sagenocrinites* is cylindrical and composed of numerous, very thin columnals. Its crown is very broad and oval in shape. The polygonal plates of the lower arms display clear growth lines and form part of the cup. The free parts of the arms are short and made up of many small plates.

HABITAT The very compact form of this species suggests that it had adapted to relatively turbulent sea-bed conditions.

stem of very thin columnals

Sagenocrinites expansus (Phillips); Wenlock Limestone; Middle Silurian; UK.

short, free arms

broad crown

arm bases forming part of cup

Typical cup diameter 1 in (2.5 cm)

Range: Middle Silurian	Distribution: Europe, N. America	Occurrence: ◉

Group: UNCLASSIFIED	Subgroup: UNCLASSIFIED	Informal name: Crinoidal limestone

Clifton Black Rock

This striking black limestone contains a length of crinoid stem and many small crinoid ossicles in the background. The structure of the stem is well shown: it is made up of short columnals of even height and has a large central cavity running along its length, now full of the limestone.

HABITAT Crinoidal limestone forms from the detritus washed from a reef.
REMARK The black color is due to bitumen.

crinoid stem

large central cavity

Clifton Black Rock; Carboniferous Limestone; Carboniferous; UK.

Typical cup diameter 1½ in (4 cm)

Range: Carboniferous	Distribution: UK	Occurrence: ◉

Group: MONOBATHRIDA	Subgroup: ACTINOCRINITIDAE	Informal name: Sea lily

Actinocrinites

This well-preserved cup is typical of many crinoids of the subclass Camerata. The large, rigid, globular cup is made up of many polygonal plates, clearly outlined on this specimen. These plates are arranged into several rings at the base, above the point of attachment to the stem. Above this point, the plates of the lower arms form a more irregular mosaic pattern, passing out to the five projecting stumps of the arms. The domed upper surface of the cup consists of fairly large plates. The stem is circular and includes many very short columnals. In the Early Carboniferous sea, current action could cause such large drifts of crinoidal debris that they became a major constituent of the limestone. A separate artificial classification has been developed for isolated stem segments.

HABITAT *Actinocrinites* lived on reefs in deep water, anchored to the sea bed by a rootlike structure. Its short, simple arms trapped small food particles.

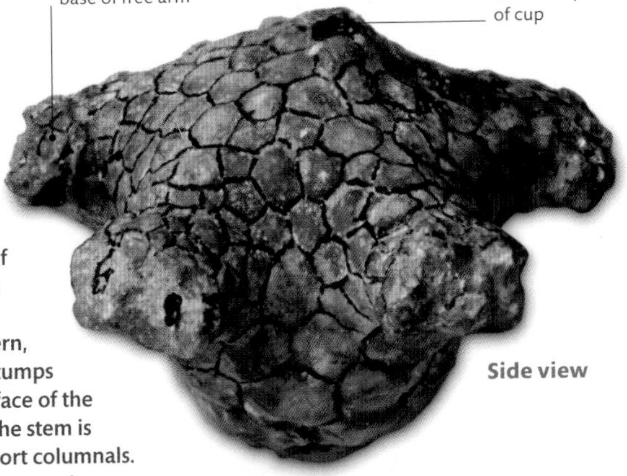

base of free arm

domed top of cup

Side view

Actinocrinites parkinsoni
Wright; Carboniferous Limestone; Early Carboniferous; UK.

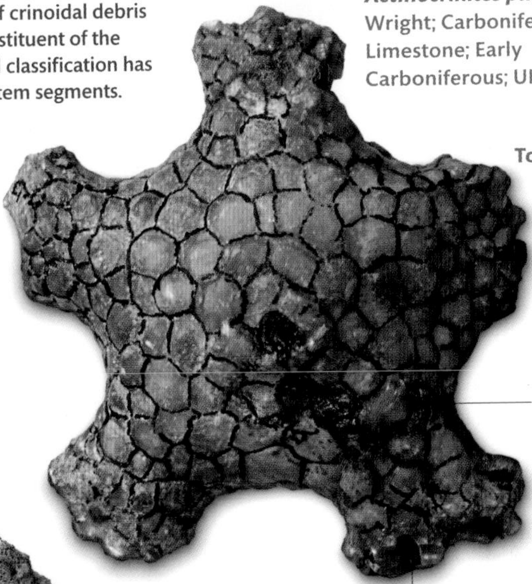

Top view

curved surface between arms

Crinoidal limestone

polygonal plates

stem segments

Typical cup diameter
1¹/₂ in (4 cm)

Range: Carboniferous	Distribution: Worldwide	Occurrence: ●●●●●

Group: ROVEACRINIDA	Subgroup: SACCOCOMIDAE	Informal name: Feather star

Saccocoma

This is a delicately constructed, stemless crinoid, with a small, globular cup. The arms branch once to give a total of 10. The lower parts of the arms have small, winglike extensions; the upper parts have long, narrow side branches.

HABITAT *Saccocoma* was free-living and probably able to tolerate variations in salinity.

REMARK This species is one of the most common fossils in the Solnhofen Limestone.

globular cup

branched arms

Typical cup diameter
³/₄ in (2 cm)

Saccocoma tenella
(Goldfuss); Solnhofen Limestone;
Late Jurassic; Germany.

Range: Late Jurassic	Distribution: Worldwide	Occurrence: ◉◉◉◉◉

Group: CYRTOCRINIDA	Subgroup: SCLEROCRINIDAE	Informal name: Sea lily

Torynocrinus

This curiously shaped crinoid has arms set at right angles to its cup, which is shaped like a spoon. The stemlike extension is actually formed from elongated cup places and attached directly to a holdfast.

HABITAT *Torynocrinus* lived attached to pebbles by a holdfast and inhabited shallow, turbulent waters.

REMARK The cyrtocrinids are squat, compact crinoids, locally common in Cretaceous rocks.

end of cup

arm attachment point

point of attachment to root

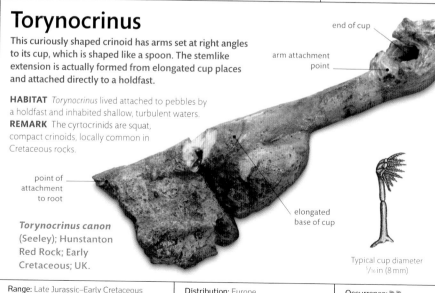

Torynocrinus canon
(Seeley); Hunstanton
Red Rock; Early
Cretaceous; UK.

elongated base of cup

Typical cup diameter
⁵/₁₆ in (8 mm)

Range: Late Jurassic–Early Cretaceous	Distribution: Europe	Occurrence: ◉◉

Group: COMATULIDA	Subgroup: PTEROCOMIDAE	Informal name: Feather star

Pterocoma

This comatulid or feather scar has a small cup, which bears 10 long, featherlike arms of equal length. Each of these has a row of little branches (pinnules) on either side. It possessed a stem only in the earliest stages of development.

HABITAT *Pterocoma* was a free-swimming crinoid, advancing with graceful movements of the arms, like its living relatives.
REMARK Comatulids are rarely found as entire specimens.

long, featherlike arms

Pterocoma pinnata (Schlotheim); Solnhofen Limestone; Late Jurassic; Germany.

small, centrally placed cup

Typical cup diameter
⅝ in (1.5 cm)

Range: Late Jurassic–Late Cretaceous	Distribution: Europe, Asia	Occurrence: ◉

Group: ENCRINIDA	Subgroup: ENCRINIDAE	Informal name: Sea lily

Encrinus

crown, with 10 tightly closed arms

long, round stem

The long, cylindrical stem of this crinoid includes regularly spaced, swollen columnals. The crown is squat and compact; the plates are large and swollen. Each arm branches once at the base to give a total of 10 short, sturdy arms.

HABITAT *Encrinus* lived in shallow seas.
REMARK This creature could splay its arms to form a feeding fan. Plankton, entrapped in the fan, were conveyed to the mouth along the grooves that lie inside the arms. When threatened, the arms closed tightly together in a defensive movement.

Typical cup diameter
1 in (2.5 cm)

Encrinus liliiformis Lamarck; Muschelkalk; Middle Triassic; Germany.

Range: Middle Triassic	Distribution: Europe	Occurrence: ◉◉

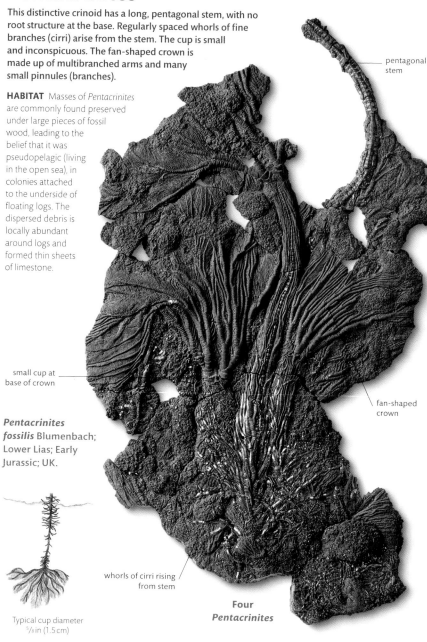

Group: ISOCRINIDA	Subgroup: PENTACRITINIDAE	Informal name: Sea lily

Pentacrinites

This distinctive crinoid has a long, pentagonal stem, with no root structure at the base. Regularly spaced whorls of fine branches (cirri) arise from the stem. The cup is small and inconspicuous. The fan-shaped crown is made up of multibranched arms and many small pinnules (branches).

HABITAT Masses of *Pentacrinites* are commonly found preserved under large pieces of fossil wood, leading to the belief that it was pseudopelagic (living in the open sea), in colonies attached to the underside of floating logs. The dispersed debris is locally abundant around logs and formed thin sheets of limestone.

pentagonal stem

small cup at base of crown

fan-shaped crown

Pentacrinites fossilis Blumenbach; Lower Lias; Early Jurassic; UK.

whorls of cirri rising from stem

Four Pentacrinites

Typical cup diameter 5/8 in (1.5 cm)

Range: Jurassic	Distribution: Europe, N. America	Occurrence: ◑◑

| Group: MILLERICRINIDA | Subgroup: APIOCRINITIDAE | Informal name: Sea lily |

Apiocrinites

This stoutly constructed crinoid has a narrow, cylindrical, tapering stem with a conical, irregularly shaped root cemented at the base. The bulbous cup is made up of enlarged columnals, two cycles of cup places, and the bases of the arms. The arms branch once, symmetrically, to give an elegant, fan-shaped crown with 10 arms in total.

HABITAT *Apiocrinites* lived attached to hard, current-swept pavements on the Jurassic sea floor. It filtered food from the water with its arms.
REMARK Remains of this species are found most commonly as separate discoidal cup plates or columnals. The surface of the stem and the holdfast are often colonized by bryozoa and serpulid worms.

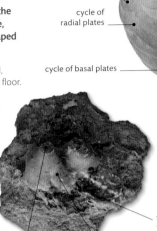

broken base of arms

large, bulbous cup made of separate plates

cycle of radial plates

cycle of basal plates

Apiocrinites elegans (Defrance); Bradford Clay; Middle Jurassic; UK.

stem cemented to sea floor

Typical cup diameter 1¼ in (3 cm)

column base attachment

conical root

| Range: Jurassic–Cretaceous | Distribution: Europe, Africa, N. America | Occurrence: ◉◉◉◉ |

| Group: UINTACRINIDA | Subgroup: MARSUPITIDAE | Informal name: Sea lily |

Marsupites

The cup is large and globular, comprising 11 large, thin, polygonal plates, which have a radially arranged ornament. There is no stem, but in its place sits a large, pentagonal plate. The arms are long and very narrow and branch to give a total of 10.

HABITAT The adult *Marsupites* nestled in chalky mud on the sea floor and strained food using its arms. The larval form was planktonic.
REMARK The large, separate cup plates are distinctive fossils and are used as zonal markers in western Europe.

branching arms

Marsupites testudinarius (Schlotheim); Upper Chalk; Late Cretaceous; UK.

large cup plates

central pentagonal plate, without stem

Typical cup diameter 1²/₅ in (3.5 cm)

| Range: Late Cretaceous | Distribution: Worldwide | Occurrence: ◉◉◉◉ |

Group: UINTACRINIDA	Subgroup: UINTACRINIDAE	Informal name: Sea lily

Uintacrinus

This crinoid has no stem but has a very large and globular cup composed of many small, polygonal plates. The bases of the arms are incorporated in the cup, and the arms themselves are very narrow and greatly elongated, reaching up to 50 in (1.25 m).

Uintacrinus socialis
Grinnell; Niobrara Chalk; Late Cretaceous; US.

HABITAT Crinoids of this genus lived in soft, muddy areas of the sea floor, using their very long arms to catch plankton.

REMARK Fossilized remains of *Uintacrinus* are usually found in the form of separate plates.

Typical cup diameter 1³⁄₄ in (4.5 cm)

large, globular cup

base of long, narrow arms

fragments of arms

Range: Late Cretaceous	Distribution: Worldwide	Occurrence: ◉◉◉◉◉

Group: ISOCRINIDA	Subgroup: ISOCRINIDAE	Informal name: Feather star

Isocrinus

Isolated crinoid stem segments consisting of up to a dozen columnal ossicles are a common occurrence in almost every Jurassic marine sediment. In some horizons, they are sufficiently abundant to form thin limestones. In section, isocrinid columnals are shaped like five-pointed stars, as opposed to the pentagonal stems of other related crinoids (see p.172). They also possess whorls of cirrae every 10–20 ossicles. The facets of the ossicles bear distinctive petaloid markings resembling starfish.

HABITAT *Isocrinus* was a suspension feeder, living in a shallow shelf sea, attached to the sea floor by a rootlike holdfast.

columnal

cirrus attachment scar

Typical cup diameter ⁵⁄₈ in (1.5 cm)

Isocrinus basaltiformis (Miller); Early Jurassic; UK.

petaloid marking

ossicle

Range: Triassic–Miocene	Distribution: Worldwide	Occurrence: ◉◉◉◉◉

ECHINOIDS

THE ECHINOIDS possess a rigid, globular skeleton (test) made up of columns of thin, calcite plates (ambulacrals and interambulacrals). The plates known as ambulacrals have small pores for tube feet. All plates have swollen tubercles for the ball-and-socket articulation of spines, which are used for defense and sometimes for walking. Regular echinoids, which forage on the sea bed, show radial symmetry; irregular echinoids, which usually burrow in soft sea beds, show bilateral symmetry.

Group: ECHINOCYSTITOIDA	Subgroup: PALAECHINIDAE	Informal name: Sea urchin

Melonechinus

The large, nearly spherical test is formed of many thick, polygonal plates. Both ambulacrals and interambulacrals are made up of numerous plate columns. The midline of the ambulacrals forms a distinct ridge.

HABITAT *Melonechinus* lived in areas between reefs in the Carboniferous sea.

Typical diameter 3 in (8 cm)

numerous columns of polygonal plates

Upper surface of test

ridged ambulacral columns

Melonechinus multipora (Owen & Norwood); Early Carboniferous; US.

Range: Early Carboniferous	Distribution: Worldwide	Occurrence: ◉◉

Group: CIDAROIDA	Subgroup: POLYCIDARIDAE	Informal name: Sea urchin

Plegiocidaris

The circular, slightly flattened test comprises five paired, narrow ambulacral plate columns and five paired, broad interambulacra made up of large plates with prominent tubercles. The hole on the top is the apical disk.

HABITAT *Plegiocidaris* lived on rocky sea floors, grazing on algae.
REMARK Large spines were attached to the tubercles in life.

Typical diameter 1¹/₂ in (4 cm)

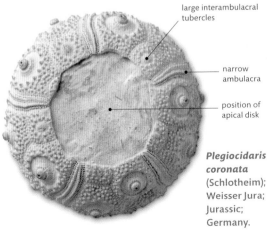

large interambulacral tubercles

narrow ambulacra

position of apical disk

Plegiocidaris coronata (Schlotheim); Weisser Jura; Jurassic; Germany.

Range: Jurassic	Distribution: Europe	Occurrence: ◉◉◉

Group: CIDAROIDA	Subgroup: ARCHAEOCIDARIDAE	Informal name: Sea urchin

Archaeocidaris

The large test, which retains many spines on its surface, is crushed flat. The paired ambulacral columns are narrow, and the broad interambulacra are made up of four columns of plates, each of which carries a single, centrally placed tubercle. A long, narrow spine articulates with each large tubercle. Short spines form a feltlike covering over much of the test. The teeth and jaws are present on the underside but are dissociated.

HABITAT The genus was probably an omnivorous browser living on the open sea floor, protected by its long spines.

scattered elements of jaw system

long, narrow spine

large tubercle

small, fine spines

Lower surface of test

position of mouth

Archaeocidaris whatleyensis
Lewis & Ensom; Carboniferous Limestone; Early Carboniferous; UK.

Upper surface of test

ambulacral column

smooth spines

feltlike covering of fine spines

Typical diameter 3 in (8 cm)

position of apical disk

Range: Early Carboniferous–Permian	Distribution: Worldwide	Occurrence: ◉◉

| Group: CIDAROIDA | Subgroup: PSYCHOCIDARIDAE | Informal name: Sea urchin |

Tylocidaris

The small test is circular in outline and slightly flattened. The 10 columns of large interambulacral tubercles carry massive, club-shaped defensive spines. The central aperture on the base housed the jaw mechanism (known as Aristotle's Lantern) in life.

HABITAT *Tylocidaris* lived as an omnivorous grazer on shells and sponges on the Chalk sea floor.
REMARK Isolated spines and fragments of the test are quite common Chalk fossils.

Typical diameter 1¼ in (3 cm)

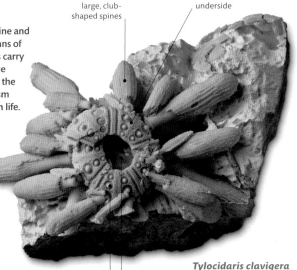

large, club-shaped spines

underside

large interambulacral tubercles

position of mouth opening

Tylocidaris clavigera (König); Upper Chalk; Late Cretaceous; UK.

| Range: Late Cretaceous–Eocene | Distribution: Europe, N. America | Occurrence: ●●●● |

| Group: CIDAROIDA | Subgroup: CIDARIDAE | Informal name: Sea urchin |

Temnocidaris

The nearly spherical test is made up of five paired columns of large interambulacral plates, each with a large conspicuous tubercle. The five paired ambulacral columns are narrow and sinuous. The large, spindle-shaped spines attached to the large interambulacral tubercles had a rough, thorny surface in life.

HABITAT *Temnocidaris* lived on the Chalk sea floor. It was an omnivorous scavenger, using its sharp teeth to rasp food.
REMARK *Temnocidaris* is very rarely preserved in its entirety. The apical disk is missing in this specimen.

Temnocidaris sceptrifera (Mantell); Upper Chalk; Late Cretaceous; UK.

narrow ambulacrum

large interambulacral tubercle

position of apical disk

large primary spine

Typical diameter 1¾ in (4.5 cm)

| Range: Late Cretaceous | Distribution: Europe | Occurrence: ●●● |

Group: SALENIOIDA	Subgroup: SALENIIDAE	Informal name: Sea urchin

Trisalenia

The circular flattened test of *Trisalenia* is capped by a very large, flat apical disk made up of two rings of plates, plus a single central plate and a hole in which the anus was situated in life. The large genital plates carry pores through which eggs and sperm were released. The large interambulacral tubercles carried smooth, pointed spines.

HABITAT *Trisalenia* lived among large granite boulders on the rocky shoreline that fringed part of southern Sweden during the Late Cretaceous.
REMARK The family Saleniidae occurs commonly in marine Cretaceous sediments.

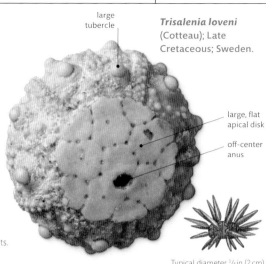

Trisalenia loveni (Cotteau); Late Cretaceous; Sweden.

large tubercle

large, flat apical disk

off-center anus

Typical diameter ³⁄₄ in (2 cm)

Range: Late Cretaceous	Distribution: Sweden	Occurrence: ●●

Group: HEMICIDAROIDA	Subgroup: HEMICIDARIDAE	Informal name: Sea urchin

Hemicidaris

The test of *Hemicidaris* has a rounded outline from above and a rather small apical disk. The 10 columns of interambulacral plates each carry a very prominent tubercle bearing long, smooth, tapering spines. The ambulacra are narrow.

HABITAT *Hemicidaris* lived among rocky outcrops on the sea bed, held by the sticky tube feet on its underside. It grazed hard surfaces using its five sharp teeth.
REMARK This well-preserved specimen was buried rapidly by sediment.

long, tapering, smooth spines

large tubercles

Typical diameter 1¹⁄₄ in (3 cm)

smaller, short spines

apical disk

Hemicidaris intermedia (Fleming); Coralline Limestone; Late Jurassic; UK.

Range: Middle Jurassic–Early Cretaceous	Distribution: Worldwide	Occurrence: ●●

| Group: PHYMOSOMATOIDA | Subgroup: PHYMOSOMATIDAE | Informal name: Sea urchin |

Phymosoma

The test is circular and flattened. The areas occupied by both the apical disk and the membrane around the mouth are very broad. The ambulacral and interambulacral tubercles are about the same size and in life bore tapering, smooth, cylindrical or flattened spines. Isolated spines are common Chalk fossils.

HABITAT Phymosoma lived on the Chalk sea floor, grazing on hard surfaces to obtain algae, sponges, and other soft organisms.

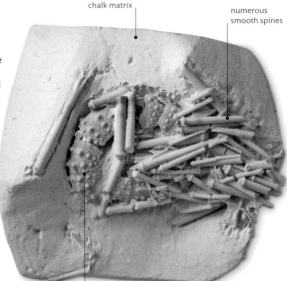

chalk matrix

numerous smooth spines

test with many tubercles

Phymosoma koenigi (Mantell); Upper Chalk; Late Cretaceous; UK.

Typical diameter 1¼ in (3 cm)

| Range: Late Jurassic–Eocene | Distribution: Worldwide | Occurrence: ◉◉◉ |

| Group: ARBACIOIDA | Subgroup: ARBACIOIDAE | Informal name: Sea urchin |

Coleopleurus

The small test has a pentagonal outline and is hemispherical in profile. The ambulacrals are broad and swollen and carry large tubercles. The interambulacrals are smooth and devoid of tubercles on the upper part of the test. The apical disk has five large genital pores and a central opening in which the anus was enclosed.

HABITAT This genus lived on hard, rocky substrates in shallow water.

REMARK The large, basally situated ambulacral tubercles imply that *Coleopleurus* had stout spines in life. It is probable that these spines were used for stability rather than for protection.

apical disk

genital pores

large tubercles on ambulacra

Coleopleurus paucituberculatus (Gregory); Morgan Limestone; Miocene; South Australia.

smooth interambulacra

Typical diameter ¾ in (2 cm)

| Range: Miocene | Distribution: Australia | Occurrence: ◉◉ |

Group: ECHINONEOIDA	Subgroup: CONULIDAE	Informal name: Sea urchin

Conulus

The test is rounded or slightly pentagonal when viewed from above and conical in side aspect. The base is flattened or concave. The mouth was positioned centrally on the base, the anus on the edge of the base. The surface of the test bears very small tubercles, which in life carried a feltlike covering of small spines.

HABITAT *Conulus* plowed through the soft mud of the sea floor. In certain thin stratigraphic levels, tests are very common.

Conulus albogalerus
Leske; Upper Chalk;
Late Cretaceous; UK.

Side view

small apical disk

ambulacral plates with pores

interambulacral plates

hole enclosing mouth

hole enclosing anus

Basal view

Typical diameter
1³/₈ in (3.5 cm)

Range: Late Cretaceous–Paleocene	Distribution: Worldwide	Occurrence: ◉◉◉

Group: CLYPEASTEROIDA	Subgroup: CLYPEASTERIDAE	Informal name: Sand dollar

Clypeaster

The large, thick test of *Clypeaster* has a rounded, pentagonal outline and a low, conical profile. On the upper surface, the ambulacra are expanded to form distinctive petals, which carry many elongated pores. The small, circular mouth is surrounded by jaws and lies at the center of the depressed base. The anus also lies on the lower surface near the posterior. The surface of the test is covered with small tubercles, which in life bear short spines.

HABITAT *Clypeaster* is a typical sand dweller, living partly or completely submerged in sand in shallow water, feeding on sediment. It is a "living fossil" and is found today in shallow, tropical seas.

Clypeaster aegyptiacu
Michelin; Miocene;
Saudi Arabia.

large petals use for respiration

tiny tubercles

Typical diameter
4³/₄ in (12 cm)

Range: Late Eocene–Recent	Distribution: Worldwide	Occurrence: ◉◉◉◉

Group: CLYPEASTEROIDA	Subgroup: MELLITIDAE	Informal name: Sand dollar

Encope

Upper surface of test

The sand dollar, *Encope*, is very flat and thin in profile. The five petals are well developed but smaller than in *Clypeaster* (see opposite). Each of the five ambulacra is perforated by an oval hole (a lunule) near the margin, which passes right through the test. A single, larger sixth hole is present in the posterior ambulacrum.

HABITAT *Encope* lives buried in sand in shallow tropical water, often in large numbers. The lunules aid the animal in feeding on fine sediment.

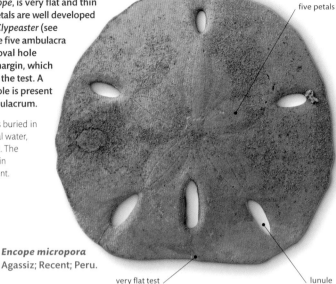

five petals

Encope micropora
Agassiz; Recent; Peru.

very flat test

lunule

Typical diameter
3¹/₂ in (9 cm)

Range: Miocene–Recent	Distribution: N. & S. America	Occurrence: ◉◉◉

Group: CLYPEASTEROIDA	Subgroup: ROTULIDAE	Informal name: Sand dollar

Heliophora

five ambulacral petals

posterior indentations

A very distinctive small African sand dollar. The test has a variable number of indentations along the posterior margin forming fragile, fingerlike projections.

HABITAT *Heliophora* lives buried just below the sediment surface in coastal lagoons and estuaries.
REMARK This species is locally abundant along the west African coast.

interambulacral plates

Heliophora orbiculus
(Linnaeus); Late Miocene; Morocco.

four genital pores

Typical diameter
1³/₈ in (3.5 cm)

Range: Miocene–Recent	Distribution: Africa	Occurrence: ◉◉◉◉

Group: CASSIDULOIDA	Subgroup: FAUJASIIDAE	Informal name: Sea urchin

Hardouinia

The large test of *Hardouinia* is conical in profile with a flat base. The petals, which in life bore tube feet specialized for respiration, are short and broad and very conspicuous. The interambulacral plates are low and broad. The mouth lies centrally on the lower surface and has five prominent projections, giving it a stellate outline. The anus is inset on the posterior side of the upper test. The surface originally carried short, fine spines.

HABITAT This genus lived partially buried in sandy sediment, into which it burrowed with the aid of the short spines. It swallowed the sand that it lived in to obtain nourishment.

REMARK *Hardouinia* is common in sandy limestone faces in the eastern US. Juvenile specimens are much less conical in shape.

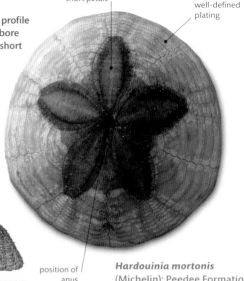

broad, short petals

well-defined plating

position of anus

Hardouinia mortonis (Michelin); Peedee Formation; Late Cretaceous; US.

Typical diameter 1⅜ in (3.5 cm)

Range: Late Cretaceous	Distribution: N. America	Occurrence: ●●●

Group: HOLASTEROIDA	Subgroup: HOLASTERIDAE	Informal name: Sea urchin

Echinocorys

The elongated test is large and conical in profile, often with a flattened top. It has a covering of small tubercles. The base is flat and the oval mouth is positioned near the anterior border. In life, the spine cover was fine and short. It has a large, elongated apical disk and long petals.

HABITAT This genus lived partially buried in mud or sand. It plowed through this to find nutritious particles of sediment.

REMARK *Echinocorys* is one of the most common of the Late Cretaceous Chalk echinoids. The top of the test is often broken. This is probably due to the attack of predators, such as fish.

posterior end

large, elongated apical disk

long petals

anterior end

Echinocorys scutata Leske; Upper Chalk; Late Cretaceous; UK.

Typical diameter 2¾ in (7 cm)

Range: Late Cretaceous–Paleocene	Distribution: Worldwide	Occurrence: ●●●●●

| Group: SPATANGOIDA | Subgroup: HEMIASTERIDAE | Informal name: Heart urchin |

Hemiaster

Heart-shaped in outline and oval in profile, the test is covered in small, fine tubercles. The mouth is positioned on the flat base near the anterior margin and the anus is positioned on the posterior. The well-developed petals are set in grooves. The anterior margin is notched.

HABITAT *Hemiaster* lives deeply buried in muddy sediment, feeding on mud, which it channels to its mouth via the anterior notch.
REMARK *Hemiaster* was abundant in muddy sediments of the Cretaceous.

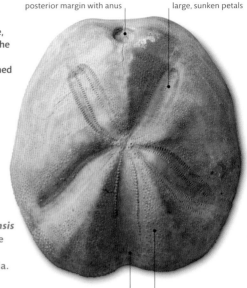

posterior margin with anus

large, sunken petals

anterior notch of test

fine tubercles

Hemiaster batnensis
Coquand; Fahdene
Formation; Late
Cretaceous; Tunisia.

Typical diameter
1½ in (4 cm)

| Range: Cretaceous–Recent | Distribution: Worldwide | Occurrence: ⊛⊛⊛ |

| Group: SPATANGOIDA | Subgroup: SCHIZASTERIDAE | Informal name: Heart urchin |

Linthia

Linthia was a rather small, compact heart urchin with a very strong, deep notch on the anterior margin and conspicuous petals with large pores. Four large genital pores on the apical disk were used for the discharge of eggs and sperm.

HABITAT This genus lived as a burrower in soft sediments.
REMARK Weathering of the test surface of this particular specimen has revealed the boundaries between plates in places.

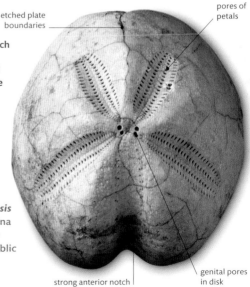

etched plate boundaries

pores of petals

strong anterior notch

genital pores in disk

Linthia sudanensis
Bather; Kalanbaina
Formation; Early
Paleocene; Republic
of Niger.

Typical diameter
2¾ in (7 cm)

| Range: Paleocene | Distribution: Northern Africa | Occurrence: ⊛⊛ |

Group: SPATANGOIDA	Subgroup: SCHIZASTERIDAE	Informal name: Heart urchin

Schizaster

Block with complete and broken heart urchins

The anterior petal in the heart urchin, *Schizaster*, is deeply inset and enlarged. The pores, which in life bear respiratory tube feet, are very conspicuous on the petals.

HABITAT *Schizaster* lives deeply buried in muds. It uses its long tube feet as a funnel-like connection to the sea bed. *Schizaster* feeds on fine particles of sediment.

REMARK This group of entire and broken individuals probably represents a storm-swept residue—the result of waves scouring deeply buried urchins from their burrows. Most specimens are preserved as internal molds, as the shell usually breaks away.

well-developed petals

clay matrix

pores of respiratory tube feet

Schizaster branderianus
Forbes; Barton Beds;
Middle Eocene; UK.

large anterior petal

Typical diameter
1 in (2.5 cm)

Range: Eocene–Recent	Distribution: Worldwide	Occurrence: ●●●●

| Group: SPATANGOIDA | Subgroup: MICRASTERIDAE | Informal name: Heart urchin |

Micraster

The Chalk heart urchin, *Micraster*, is pointed posteriorly and has a strong anterior notch. The five petals are narrow and fairly short.

HABITAT This genus lived buried in the soft mud of the Chalk Sea floor and channeled a stream of mud into its mouth via the strong anterior notch.

REMARK This species is one of the more advanced forms, occurring in the higher beds of the Chalk.

Micraster coranguinum (Leske); Upper Chalk; Late Cretaceous; UK.

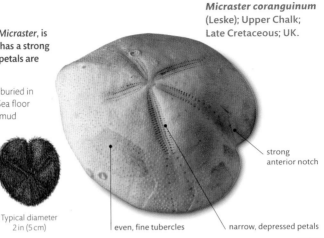

Typical diameter 2 in (5 cm)

strong anterior notch

even, fine tubercles

narrow, depressed petals

| Range: Late Cretaceous–Paleocene | Distribution: Worldwide | Occurrence: ◉◉◉◉ |

| Group: SPATANGOIDA | Subgroup: LOVENIIDAE | Informal name: Heart urchin |

Lovenia

Upper surface

posterior

This heart urchin has an elongated, flattened test with a flat base. The petals are shallow and taper toward the margin of the test. In addition to the fine covering of tubercles, large, deeply inset tubercles are present on both sides of the test. These carry long, curved, protective spines.

HABITAT *Lovenia* is a shallow burrower in sand, usually occurring in inshore marine deposits.

REMARK The few large tubercles on the upper surface of this heart urchin make it very distinctive.

anus

large tubercles

anterior notch

flat base

Lower surface

Lovenia forbesi (Woods); Morgan Limestone; Miocene; Australia.

crescent-shaped mouth

anterior

Typical diameter 1 in (2.5 cm)

| Range: Late Eocene–Recent | Distribution: Worldwide | Occurrence: ◉◉◉ |

ASTEROIDS

THE ASTEROIDS, which include many of the species popularly called starfish, are common marine animals and have a long history extending back into the Ordovician. Most have five arms, although some species have more. The mouth is centrally placed on the underside, and five ambulacral grooves, floored by ambulacral and adambulacral ossicles, run along the midline of each arm. The grooves house the soft, muscular tube feet, which are used for walking, burrowing, and manipulating prey. Asteroids are rarely preserved as complete specimens.

| Group: UNCLASSIFIED | Subgroup: PALASTERICIDAE | Informal name: Starfish |

Palastericus

This starfish has five short, straplike arms. All the ossicles, except those of the ambulacral groove and mouth, are small and inconspicuous. The body is covered with short, fine spines.

HABITAT *Palastericus* lived by ingesting sediment and small marine animals.
REMARK The specimen shown here has been beautifully preserved in iron pyrites.

Palastericus devonicus Sturtz; Hunruckschiefer; Late Devonian; Germany.

small ossicles of upper surface

Typical diameter 5 in (12 cm)

ridge formed by ossicles of ambulacral groove

| Range: Late Devonian | Distribution: Germany | Occurrence: ◉ |

| Group: PAXILLOSIDA | Subgroup: ASTROPECTINIDAE | Informal name: Starfish |

Pentasteria

The five arms are quite narrow and the disk is small. Large marginal ossicles form a broad, well-defined border to the asteroid. The upper surface of the disk is occupied by small plates through which the large mouth ossicles can be clearly seen.

HABITAT *Pentasteria* probably lived as a shallow burrower in sand.

Pentasteria cotteswoldiae (Buckman); Stonesfield Slate; Middle Jurassic; UK.

border of large marginal ossicles

Typical diameter 4 in (10 cm)

| Range: Jurassic–Early Cretaceous | Distribution: Europe | Occurrence: ◉ ◉ |

Group: STENURIDA	Subgroup: STENASTERIDAE	Informal name: Starfish

Stenaster

This small and enigmatic form has been variously classified as an ophiuroid and an asteroid. The five arms are short and the disk is rather broad in form. The broad ambulacral ossicles extend across the width of the arm.

HABITAT This was a marine-dwelling animal.
REMARK The Starfish Bed at Girvan in Scotland, UK, formerly yielded a large number of beautifully preserved starfish and other fossils. These starfish were present as natural molds, where the calcite skeletons had dissolved. A storm was probably responsible for the rapid deposition of sand that smothered the animals. *Stenaster* is a starfish with poorly understood relationships to other asteroids.

Upper surface of starfish

Stenaster obtusus (Forbes); Drummock Group; Ordovician; UK.

Lower surface of starfish

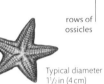

rows of ossicles

Typical diameter 1½ in (4 cm)

Range: Ordovician	Distribution: Europe, N. Africa	Occurrence: ◉◉

Group: VELATIDA	Subgroup: TROPIDASTERIDAE	Informal name: Starfish

Tropidaster

Tropidaster pectinatus Forbes; Middle Lias; Early Jurassic; UK.

This small form has five blunt, petal-shaped arms and very large, conspicuous mouth places. The ambulacral groove is widely open, and broad adambulacral ossicles make up the underside of the arm. In life, these each bore a row of prominent spines.

HABITAT This starfish dwelt in shallow marine water.
REMARK The photograph shows two specimens, a juvenile and a fully grown adult, both displaying the underside.

wide ambulacral groove

Typical diameter 1 in (2.5 cm)

underside of arm showing ambulacral ossicles

Range: Early Jurassic	Distribution: UK	Occurrence: ◉

Group: VALVATIDA	Subgroup: GONIASTERIDAE	Informal name: Starfish

Metopaster

The arms are very short and the disk is large.
(The body resembles a pentagonal cookie.) The
marginal ossicles are large, few in number, and
conspicuous; in life, they bore small, granular
spines. The marginals, which form the tips of
the arms, are elongated and triangular.
The disk was covered with small,
polygonal ossicles.

HABITAT Like its living relatives,
Metopaster probably lived as a sediment
feeder or scavenger on the Chalk Sea floor.
REMARK Marginal ossicles of this genus
are common fossils.

small
ossicles

pentagonal
form

large marginal
ossicles form
broad border

*Metopaster
parkinsoni*
(Forbes); Upper Chalk;
Late Cretaceous; UK.

Typical diameter
2 in (5 cm)

Range: Late Cretaceous	Distribution: Europe	Occurrence: ◉◉◉

Group: VALVATIDA	Subgroup: STAURANDERASTERIDAE	Informal name: Starfish

Stauranderaster

This starfish has long, narrow arms and a medium-sized
disk, which was originally dome-shaped but has collapsed
in this specimen. This disk is made up of fairly large,
rounded ossicles. The large marginal ossicles of the
arms overlap, allowing greater flexibility.

HABITAT This genus was probably
a surface dweller on the Chalk
Sea floor and either
fed on sediment
or scavenged
dead material.

Stauranderaster coronatus
(Forbes); Lower Chalk; Late
Cretaceous; UK.

long,
narrow
arms

soft chalk
matrix

marginals overlap
to increase
flexibility

disk made of large ossicles

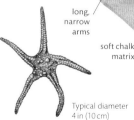

Typical diameter
4 in (10 cm)

Range: Late Cretaceous	Distribution: Europe	Occurrence: ◉

OPHIUROIDS

THE OPHIUROIDS ARE commonly called starfish or brittle stars. Closely related to echinoids, they possess five long, often fragile arms which radiate from a flat, nearly circular disk. These arms are very flexible and quite different in structure from those of asteroids. They consist of a central column of vertebrae, covered by four rows of plates. The ambulacral groove is covered in ophiuroids. Today, ophiuroids occur in vast masses on the sea floor; similar quantities are found in the fossil state.

Group: OPHIURIDA	Subgroup: OPHIODERMATIDAE	Informal name: Brittle star

Palaeocoma

The disk is small and pentagonal, with very long, sinuous arms with four rows of plates of almost equal size. It is made up of 10 large, pear-shaped plates on the upper surface and five on the lower.

HABITAT This brittle star lived on the silty sea floor.

Typical disk diameter ¾ in (2 cm)

long, flexible arms

small disk with some large plates

central mouth

Palaeocoma egertoni (Broderip); Middle Lias; Early Jurassic; UK.

Range: Early Jurassic	Distribution: Europe	Occurrence: ◉

Group: OPHIURIDA	Subgroup: APLOCOMIDAE	Informal name: Brittle star

Geocoma

The disk is small and nearly circular and the arms are long and broad at the base. The upper surface of the disk is made up of a fairly small number of large plates. The lateral plates of the arms bear short, fine spines; the upper side arm plates are very small.

HABITAT *Geocoma* lived in lime mud.

Typical disk diameter ¾ in (2 cm)

long, highly flexible arms

small, rounded central disk

fine limestone matrix

Geocoma carinata (Munster); Solnhofen Limestone; Late Jurassic; Germany.

Range: Jurassic	Distribution: Worldwide	Occurrence: ◉◉

BLASTOIDS

THE BLASTOIDS ARE a small, well-defined group of echinoderms, which in life were attached to the substrate by a thin stem, and in which the compact theca is shaped like a rosebud. The theca is made up of three circles of five plates, known as basals, radials, and deltoids. The five columns of ambulacral plates are V-shaped, bearing short brachioles used for filtering food from the water. These fossils are locally common from the Silurian to the Permian, in marine shales and limestones.

Group: SPIRACULATA	Subgroup: PENTREMITIDAE	Informal name: Blastoid

Pentremites

The small theca is biconical in form and had a very small attachment point to the stem at the lower end. The ambulacra are broad and V-shaped. The theca is made up of several very large plates and has several openings to the outside at its top end through which eggs and sperm could pass.

HABITAT This form lived attached to a hard substrate on the sea bed.
REMARK Blastoids are locally common in Carboniferous marine sediments.

openings on top of theca

large, broad ambulacra

Pentremites pyriformis J. Sowerby; Okaw Group; Early Carboniferous; US.

Typical cup diameter ¹⁄₁₆in (2mm)

Range: Early Carboniferous	Distribution: US	Occurrence: ◉◉◉

Group: SPIRACULATA	Subgroup: SCHIZOBLASTIDAE	Informal name: Blastoid

Deltoblastus

The theca is conical and tall, and the five large ambulacra are sunken into its surface. Between the ambulacra, the five areas made of large plates have a groove along the midline.

HABITAT This genus was attached to hard substrates on the sea bed.
REMARK Blastoids occur in abundance in the Permian of Timor in Indonesia, but in the rest of the world, they all but vanished in the Late Carboniferous.

Deltoblastus permicus (Wanner); Permian; Indonesia.

V-shaped regions between ambulacra

large, broad ambulacral areas

Typical cup diameter ⁵⁄₈in (1.5cm)

Range: Permian	Distribution: Indonesia	Occurrence: ◉◉◉

CYSTOIDS

CYSTOIDS SUPERFICIALLY resembled crinoids in that they were attached to the substrate by a stem and possessed a swollen theca made up of a variable number of plates. Cystoids, however, did not have true arms. Instead, they filtered food from the water by means of short, unbranched limbs called brachioles. In addition, they had distinctive triangular pore structures on the plates, which were thought to have been used for respiration. Cystoids are found, very occasionally, in Middle Ordovician to Devonian rocks.

| Group: PLEUROCYSTITIDA | Subgroup: PLEUROCYSTITIDAE | Informal name: Cystoid |

Dipleurocystis

This fossilized cystoid has a flattened, triangular-shaped theca and is broadest at the posterior margin, tapering anteriorly. The stem consists of many short, ridged columnals. The mosaic covering of small plates on the undersurface is uppermost.

HABITAT This genus lived with the undersurface lying on the sediment and with its stem coiled around any suitable attachment.

Dipleurocystis rugeri (Salter); Caradoc Series; Ordovician; UK.

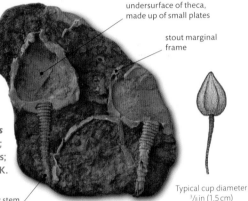

undersurface of theca, made up of small plates

stout marginal frame

tapering stem

Typical cup diameter
³/₈ in (1.5 cm)

| Range: Ordovician | Distribution: Worldwide | Occurrence: ◉◉ |

| Group: PLEUROCYSTITIDA | Subgroup: PLEUROCYSTITIDAE | Informal name: Cystoid |

Pleurocystites

This well-preserved specimen displays very clearly the large plates on the upper surface of the theca and the rather short, rapidly tapering, coiled stem. The stem is broad and flexible close to the theca and composed of alternately large and small ossicles. Two long brachioles extend on either side of the centrally placed mouth.

HABITAT *Pleurocystites* was a benthonic, filter-feeding organism.

Pleurocystites filitextus Billings; Trenton Limestone; Ordovician; Canada.

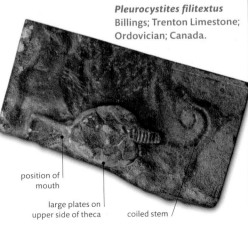

Typical cup diameter
³/₄ in (2 cm)

position of mouth

large plates on upper side of theca

coiled stem

| Range: Ordovician | Distribution: Worldwide | Occurrence: ◉◉◉◉ |

| Group: GLYPTOCYSTITIDA | Subgroup: CALLOCYSTIDAE | Informal name: Cystoid |

Lepadocrinites

Lepadocrinites had an elliptically shaped theca, which is made up of a few rather large plates, with very large, rhomb-shaped respiratory pores. The five ambulacra are long and in life bore numerous brachioles. The stem is elongated and tapering and made up of ridged columnals, which are short near the theca but are longer on the stem.

HABITAT In life, *Lepadocrinites* was attached to hard substrates on the sea floor by a root.

Lepadocrinites quadrifasciatus Pearce; Wenlock Limestone; Silurian; UK.

theca with five ambulacra

long, tapering stem

originally attached by root

Typical cup diameter 1 in (2.5 cm)

| Range: Silurian | Distribution: UK | Occurrence: ●● |

| Group: RHOMBIFERA | Subgroup: CALLOCYSTIDAE | Informal name: Cystoid |

Pseudocrinites

The theca is strongly flattened and circular in outline and has two narrow ambulacra that run around the margins and carry short, stout brachioles. The theca is made up of a few very large plates. It has one or two very large, rhomb-shaped respiratory pores. The short, tapering stem culminates in a club-shaped root.

HABITAT This genus lived as a filter feeder, attached to a hard surface, such as a shell or pebble, on the sea floor.

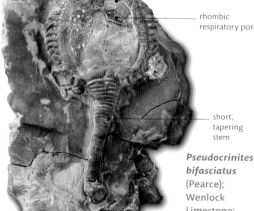

flattened theca carrying brachioles

rhombic respiratory pore

short, tapering stem

Pseudocrinites bifasciatus (Pearce); Wenlock Limestone; Silurian; UK.

Typical cup diameter 1 in (2.5 cm)

| Range: Late Silurian–Early Devonian | Distribution: Europe, N. America | Occurrence: ●● |

CARPOIDS

THE CARPOIDS, also known as Homalozoa, are a small but diverse group of invertebrates ranging from Cambrian to the Devonian. They possessed a fragile skeleton with the distinctive fine calcite structure of echinoderms but without a trace of their five-fold symmetry.

They were probably ancestral to the echinoderms, but not to the chordates as was once thought. Carpoids were free living, being able to move on the sea bed, and had a large head and a short, flexible tail. Gill slits (like those in fish) are conspicuous in some carpoid genera.

Group: CORNUTA	Subgroup: COTHURNOCYSTIDAE	Informal name: Cornute

Cothurnocystis

The head is boot-shaped and has three pointed projections. A frame of marginal plates forms the border of the head, and the central area is made up of small plates; a single row of large gill slits are also present. The tail is quite short and flexible.

HABITAT *Cothurnocystis* lived on the sea floor and was probably able to pull itself along with its tail.

Typical cup diameter 2 in (5 cm)

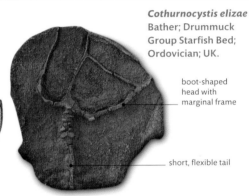

Cothurnocystis elizae Bather; Drummuck Group Starfish Bed; Ordovician; UK.

boot-shaped head with marginal frame

short, flexible tail

Range: Ordovician	Distribution: Scotland	Occurrence: ◉

Group: MITRATA	Subgroup: ANOMALOCYSTITIDAE	Informal name: Mitrate

Placocystites

The head of this mitrate is flattened and rectangular in shape. It is made up of large, thin, calcite plates, which bear fine, wavy ridges. The anterior end has two rodlike appendages which are, in fact, spines. The mouth is placed centrally.

HABITAT *Placocystites* lived on the surface of the sea bed and used its flexible tail as a lever to haul itself about.
REMARK The tail is missing in this specimen.

Typical cup diameter ¾ in (2 cm)

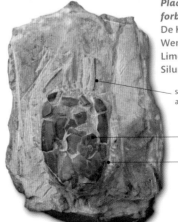

Placocystites forbesianus De Koninck; Wenlock Limestone; Silurian; UK.

spines on anterior end

head made up of thin plates

wavy ridges on surface

Range: Late Silurian	Distribution: Europe	Occurrence: ◉

VERTEBRATES

AGNATHANS

THESE EARLY, jawless fishes have an ancient lineage, represented today by the lampreys and hagfishes. Living forms have funnel-like, suctional mouths with rasping, horny teeth used for scraping flesh. Unlike jawed vertebrates, agnathan gills face inward from the gill arches. They have no paired fins. Most of the extinct forms were tadpole-shaped, swam by undulating their tails, and had thick plates and scales for armor. They first appeared in marine waters, then spread to fresh and brackish waters, dominating the Silurian and Devonian periods.

Group: PTERASPIDIFORMES	Subgroup: PTERASPIDIDAE	Informal name: Pteraspid

Pteraspis

The flattened head region of *Pteraspis* was enclosed by massive, bony plates, which may have been formed by scales fusing together. It had one middle dorsal plate, one rostral plate, and one ventral plate, plus a smaller lateral series. The plates were punctuated by sensory canals. The mouth, situated on the underside, was flanked by small plates, which may have assisted suctional bottom feeding. The eyes were small and placed along the sides. The tail, with its larger lower lobe, caused upward driving of the head when rising from the bottom. A massive bony carapace may have acted as a phosphate store during times of shortage.

HABITAT Fossil remains of *Pteraspis* are often found in marine and freshwater deposits.

Pteraspis rostrata Agassiz; Old Red Sandstone; Devonian; UK.

long rostrum

Head shield

growth line

immovable lateral plate

bony dorsal shield

dorsal spine base

Typical length 10 in (25 cm)

Range: Early Devonian	Distribution: Europe, Asia, N. America	Occurrence:

Group: CEPHALASPIDIFORMES	Subgroup: CEPHALASPIDAE	Informal name: Cephalaspid

Cephalaspis

The small freshwater fish *Cephalaspis* had a bony dorsal head shield, with overlapping scales covering the rest of its body. The eyes were situated on top, close to the midline, with the mouth on the underside. A middle dorsal area and two lateral areas of polygonal plates are thought to represent sensory areas.

HABITAT *Cephalaspis* inhabited freshwater pools or streams.

Cephalaspis whitei
Stensio; Old Red Sandstone; Devonian; UK.

sensory plate area

eye socket

cornua directed toward the back

notch

Typical length 8³/₄ in (22 cm)

Head shield

Range: Early Devonian	Distribution: Europe	Occurrence:

Group: ANASPIDIFORMES	Subgroup: BIRKENIIDAE	Informal name: Anaspid

Birkenia

The spindle-shaped body of this small fish is armored by deep, overlapping, articulated scales arranged in rows. A row of ridge scales runs along the top. The terminal mouth forms a vertical slit, surrounded by the smaller, less organized cranial scales.

HABITAT *Birkenia* was a freshwater genus.

Birkenia elegans
Traquair; Slot Burn Formation; Middle Silurian; UK.

long, spindle-shaped body

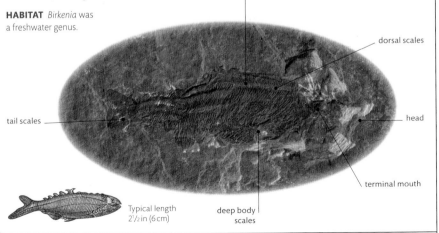

dorsal scales

tail scales

head

terminal mouth

Typical length 2¹/₂ in (6 cm)

deep body scales

Range: Middle Silurian	Distribution: Europe	Occurrence:

PLACODERMS

THE PLACODERMS FORMED a diverse group of now-extinct fishes that had simple jawbones armed with slicing plates. Typically, these fish lived on the sea bed. They had a heavily armored head shield connected to a trunk shield, which covered the anterior part of the body. The rest of the body and tail was covered by small scales. The moderate-to-large eyes were protected by a circlet of bony plates. Placoderms lived in marine and fresh water, from the Late Silurian to the Early Carboniferous. Most were moderately sized, but some grew to 20 ft (6 m) in length.

Group: ANTIARCHI	Subgroup: BOTHRIOLEPIDAE	Informal name: Antiarch

Bothriolepis

This armored fish had two dorsal and pelvic fins and a heterocercal tail. Its semicircular head shield and massive trunk shield were flattened from top to bottom. Pectoral appendages, more extended than those of *Pterichthyodes*, were probably used for balancing the body on the substrate. Paired sacs inside the trunk shield may have functioned as lungs when freshwater lakes dried up. Soft part preservation indicates the presence of a spiral intestinal valve, like that of sharks.

HABITAT *Bothriolepis* was principally a freshwater bottom feeder, but its fossil remains have been found in estuarine and marine deposits.
REMARK Species of *Bothriolepis* have been described from every continent, including Antarctica. Its success can be attributed to its heavy protective armor and its ability to adapt to a wide range of environments.

Bothriolepis canadensis
Whiteaves; Escuminac Formation; Late Devonian; Canada.

common eye sockets

semicircular head shield

long pectoral appendage

Body shield

bony trunk shield

Typical length 16 in (40 cm)

Range: Late Devonian	Distribution: Worldwide	Occurrence:

Group: ARTHRODIRA	Subgroup: COCCOSTEIDAE	Informal name: Arthrodire

Coccosteus

This predatory fish had a broad, flattened skull, with eyes placed well forward on the sides. The robust, powerful, slightly gaping jaws lacked true teeth but were armed with bony, shearing cusps, which wore down with use.

HABITAT *Coccosteus* was an active predator that inhabited shallow freshwater lakes and rivers.
REMARK Its wide geographic distribution suggests that it was able to tolerate saltwater.

Coccosteus cuspidatus
Miller; Old Red Sandstone; Devonian; UK.

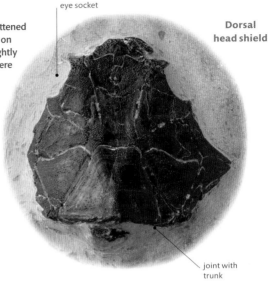

eye socket

Dorsal head shield

joint with trunk

Typical length 14 in (35 cm)

Range: Middle–Late Devonian	Distribution: Europe, N. America	Occurrence:

Group: ANTIARCHI	Subgroup: ASTEROLEPIDAE	Informal name: Antiarch

Pterichthyodes

This small fish had a pair of armored and jointed appendages, articulated with the trunk, which may have been used for bottom crawling. The relatively small head was enclosed in a bony shield, with eyes on the top and jaws on the underside, which may have been used as shovels. The massive trunk shield, with its flat undersurface, is covered with overlapping bony plates.

HABITAT *Pterichthyodes* was a bottom feeder that lived in shallow freshwater lakes.
REMARK The name *Pterichthyodes*, "winged fish," refers to its large, winglike pectoral appendages. It was originally thought that they could be used to "walk" on land, an interpretation now discounted.

small head shield

articulated appendage

massive trunk shield

Pterichthyodes milleri (Agassiz); Old Red Sandstone; Devonian; UK.

Body shield

Typical length 6 in (15 cm)

Range: Middle Devonian	Distribution: UK	Occurrence:

CHONDRICHTHYANS

CREATURES OF THIS CLASS are characterized by a cartilaginous skeleton made up of tiny, calcified prisms. Living examples include the sharks, skates, rays, and rabbitfishes. Cartilage is not usually preserved, so common fossils tend to be teeth, scales, and dorsal-fin spines. Sharks' teeth are continuously replaced from behind, being shed from the jaw margin or lost during feeding. Trace fossils (see p.42) ascribed to the group include spirally coiled fecal remains and intestines. Chondrichthyans diversified from Devonian origins to become very common in today's seas.

Group: PETALODONTIFORMES	Subgroup: PETALODONTIDAE	Informal name: Petalodont

Petalodus

The anterior teeth of this predator have a symmetrical, triangular crown, with a moderately high central cusp and pronounced cutting edge. Lateral teeth are shorter and less symmetrical.

HABITAT *Petalodus* probably inhabited coral reefs.

Anterior tooth

Petalodus acuminatus Agassiz; Yoredale Beds; Early Carboniferous; UK.

broadly triangular crown

constricted crown/ root junction

deep root

Typical length 11½ ft (3.5 m)

Range: Early Carboniferous–Permian	Distribution: Northern hemisphere	Occurrence:

Group: EUGENIODONTIFORMES	Subgroup: AGASSIZODONTIDAE	Informal name: Shark

Helicoprion

The front teeth of *Helicoprion* grew in a spiral containing up to 180 teeth, and unlike most sharks, it retained its teeth even after growing new ones. The older teeth were housed in a cavity at the junction of the lower jaws. Individual teeth consist of upright, triangular crowns over a projecting root.

HABITAT *Helicoprion* was probably a midwater marine predator.

small, juvenile teeth

Helicoprion bessonowi Karpinsky; Early Permian; Russia.

Tooth spiral

Estimated length 18 ft (5.5 m)

large, mature teeth

Range: Early Permian	Distribution: Worldwide	Occurrence:

Group: HYBODONTIFORMES | Subgroup: ACRODONTIDAE | Informal name: Hybodont shark

Acrodus

The dorsal fin of this hybodont shark was supported by an extended spine, with coarse longitudinal ridges (costae) on the side walls and a trailing edge covered with small, toothlike projections. A long basal part, inserted into the soft tissues of the back, was supported by a fin cartilage. The fin itself cut through the water and prevented rolling during swimming. The underslung mouth was armed with a battery of robust teeth for crushing its food. Coarse ridges of each individual tooth radiate from the crown center, providing additional abrasion. Anterior and posterior teeth are small; lateral teeth have greatly expanded crowns supported on robust roots.

HABITAT *Acrodus* was a slow-swimming marine shark, living close to the sea bottom on a diet of mollusks and crustaceans.

Dorsal fin spine

Acrodus nobilis
Agassiz; Lower Lias;
Early Jurassic; UK.

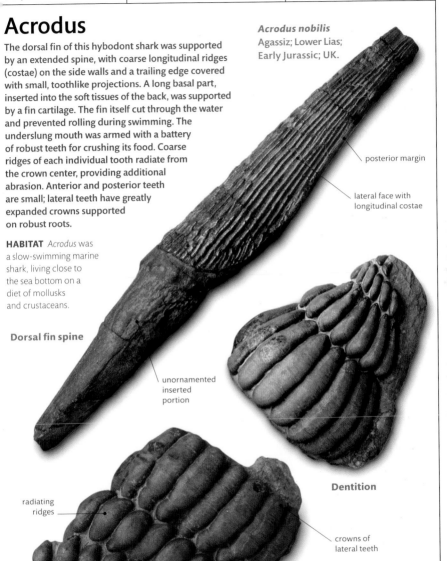

posterior margin

lateral face with longitudinal costae

unornamented inserted portion

Dentition

radiating ridges

crowns of lateral teeth

small posterior teeth

Typical length 9ft (2.7m)

Range: Triassic–Late Cretaceous | Distribution: Worldwide | Occurrence:

Group: SYNECHODONTIFORMES	Subgroup: PALEOSPINACIDAE	Informal name: Paleospinacid shark

Palidiplospinax

This was a small, probably slow-swimming shark, similar in shape to a dogfish. It had a weakly calcified braincase, with a short snout and underslung mouth containing multipointed teeth. These were ornamented by vertical ridges, with crowns displaced in relation to the tongue. The high, central cusp was flanked by up to four pairs of lower, lateral cusplets. Teeth at the front were upright and symmetrical, whereas toward the back, symmetry decreased, cusplets became shorter, and central cusps became more inclined. The side walls of the short, dorsal-fin spine were smooth and enameled, with round enamel projections at the base. Vertebrae were calcified and spool shaped, with a central perforation for the skeletal rod (notochord). Scales were simple and nongrowing.

HABITAT *Palidiplospinax* lived in shallow, marine environments, feeding on small fish and thin-shelled, bottom-dwelling invertebrates.

Palidiplospinax enniskilleni Duffin and Ward; Lower Lias; Early Jurassic; UK.

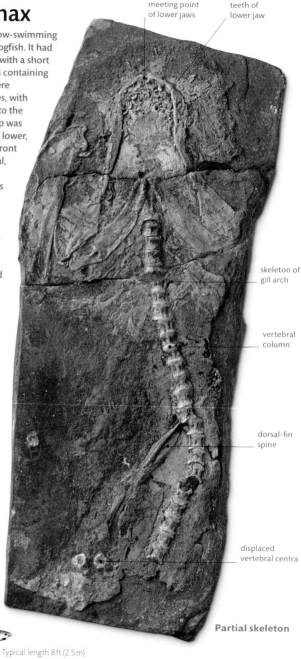

meeting point of lower jaws

teeth of lower jaw

skeleton of gill arch

vertebral column

dorsal-fin spine

displaced vertebral centra

Partial skeleton

Typical length 8 ft (2.5 m)

Range: Early Triassic–Paleocene	Distribution: Worldwide	Occurrence:

Group: HYBODONTIFORMES	Subgroup: PTYCHODONTIDAE	Informal name: Ptychodus

Ptychodus

This moderately large shark is known only from fossils of its shell-crushing teeth and possibly from calcified vertebrae. The teeth were arranged in a tightly packed battery, with massive rectangular crowns traversed by sharp, coarse ridges, giving way to round projections at the tooth margin. The square root, directly underneath the crown, was perforated by tiny, closely packed blood vessels.

HABITAT *Ptychodus* lived in shallow marine conditions, preying on thick-shelled invertebrates.

massive, rectangular, enameled crown

Ptychodus latissimus
Agassiz; Upper Chalk; Late Cretaceous; UK.

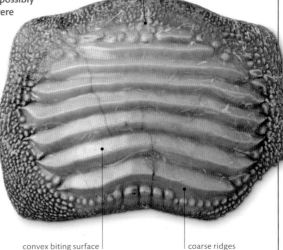

Single tooth

Typical length 10 ft (3 m)

convex biting surface

coarse ridges

Range: Late Cretaceous	Distribution: Worldwide	Occurrence:

Group: LAMNIFORMES	Subgroup: ANACORACIDAE	Informal name: Anacoracid shark

Squalicorax

Like the modern tiger shark, *Squalicorax* had triangular, flattened teeth, with finely serrated crowns. In anterior files, the teeth are upright but become increasingly inclined toward the back. The simple flat root lacks a nutritive groove.

HABITAT This shark usually inhabited shallow marine waters.

Squalicorax pristodontus
(Agassiz); Phosphate Formation; Late Cretaceous; Morocco.

polished enamel

finely serrated cutting edge

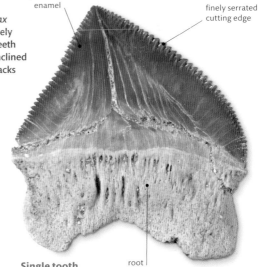

root

Single tooth

Typical length 8 ft (2.5 m)

Range: Late Cretaceous	Distribution: Worldwide	Occurrence:

| Group: LAMNIFORMES | Subgroup: ODONTASPIDAE | Informal name: Sand shark |

Striatolamia

This extinct shark is known from fossils of its teeth and calcified vertebrae. In the front teeth, the crown is narrow and tapering, becoming triangular in side and back teeth. The lingual surface (facing the tongue) is ornamented with fine raised grooves (striae), and on either side of the crown is a single cusp—conical or broad and flat. A distinct nutritive groove divides the root.

HABITAT *Striatolamia* is closely related to the modern sand shark and could tolerate low salinities.
REMARK This and similar sand shark teeth are common in many Paleocene and Eocene deposits.

root   nutritive groove

spatulate side cusp

striated crown

Striatolamia macrota (Agassiz); Barton Clay Formation; Middle Eocene; UK.

Typical length 11½ ft (3.5 m)

Upper lateral tooth

| Range: Paleocene–Eocene | Distribution: Worldwide | Occurrence: |

| Group: HEXANCHIFORMES | Subgroup: HEXANCHIDAE | Informal name: Cow shark |

Notorynchus

This is a seven-gilled shark with multicusped teeth, of which the lower teeth have crowns, comprising a principal cusp, and three to eight cusplets spreading out from the center. To the front of the principal cusp is a further series of small cusplets. The root is rectangular and flattened. Upper teeth are smaller and narrower.

HABITAT
Notorynchus lives in cool, shallow marine waters.

multicusped crown

Notorynchus kempi Ward; Barton Clay Formation; Middle Eocene; UK.

principal cusp

mesial cusplets

junction of root and crown

root

Typical length 10 ft (3 m)

Lower anterolateral tooth

| Range: Eocene–Recent | Distribution: Worldwide | Occurrence: |

| Group: LAMNIFORMES | Subgroup: LAMNIDAE | Informal name: Great white shark |

Carcharodon

The teeth have large, triangular crowns and flattened roots. Younger individuals have narrower teeth and often possess a pair of vestigial lateral cusps. The more recent species have evolved serrated crowns. Upper teeth are wider, and the more lateral teeth point backward. Lower teeth are smaller and almost symmetrical.

HABITAT Modern great white sharks are partially warm-blooded and prefer temperate rather than tropical waters. They hunt mainly inshore, living near the surface and feeding on marine mammals such as seals and dolphins.

REMARK The Carcharodon lineage first arose from Mako shark–like ancestors in the Oligocene with *Carcharodon hastalis*. In the Late Miocene of the eastern Pacific, a lightly serrated species called *C. hubbelli* evolved, which gave rise to the modern great white shark. Meanwhile, in the Atlantic Ocean, an unserrated species with wide teeth evolved, known as *C. plicatilis*. This became extinct in Late Pliocene times and was replaced globally by the great white shark, *C. carcharias*, which lives to this day, although endangered.

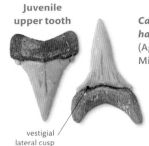

Juvenile upper tooth

Carcharodon hastalis (Agassiz); Late Miocene; Chile.

Juvenile lower tooth

vestigial lateral cusp

First upper anterior tooth

Second upper anterior tooth

Labial (outside facing) views

Second lower anterior tooth

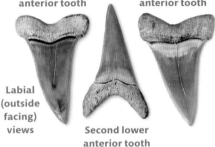

Carcharodon hastalis (Agassiz); Early Miocene; US.

First upper anterior tooth

Second upper anterior tooth

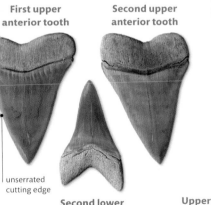

unserrated cutting edge

Second lower anterior tooth

Carcharodon plicatilis (Agassiz); Early Pliocene; US.

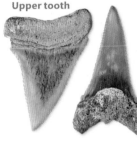

Upper tooth

Carcharodon hubbelli (Ehret et al.); Late Miocene; Chile.

Lower tooth

Upper tooth

serrated cutting edge

Carcharodon carcharias (Linnaeus); Early Pliocene; Chile.

Lower tooth

Typical tooth height 2 in (5 cm)

| Range: Late Oligocene–Recent | Distribution: Worldwide | Occurrence: |

Group: LAMNIFORMES	Subgroup: OTODONTIDAE	Informal name: Megatooth shark

Otodus

The genus *Otodus* is part of an extinct lineage of sharks whose origins lie in the middle Cretaceous. Individual species are characterized by massively constructed, broadly triangular teeth with a distinct chevron-shaped neck. Anterior teeth are relatively upright, while lateral teeth are distally inclined, more so in the upper teeth.

Otodus remains are usually restricted to isolated teeth that have been shed, but occasionally, associated dentitions and vertebral columns have been found. Between the Paleocene and early Pliocene, a series of changes take place in the dentition. The overall tooth size increases from about 2 in (5 cm) to over 6 in (15 cm). The cutting edges develop serrations, initially irregular and toward the base of the crown, then with time more even and finer, including the crown tips. The lateral cusps also become serrated, initially coarsely, then more evenly, matching the cutting edges. As the base of the crown increases in size, the lateral cusps diminish in relative size and are lost.

HABITAT *Otodus* lived in warm coastal waters and was both an apex predator and an opportunistic scavenger living on large fish, turtles, and marine mammals.

REMARK *Otodus* teeth evolve from one species to another without any sudden changes. The teeth of juvenile *Otodus* tend to resemble those of their immediate ancestors, changing shape during their lifetime. Both factors make precise identifications difficult. The last species in the lineage, *Otodus megalodon*, was the world's largest predatory fish. Its sudden extinction in middle Pliocene times, 3.6 million years ago, is thought to be due to a combination of oceanic cooling, sea-level fall, and competition from toothed whales and great white sharks.

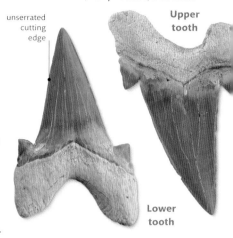

Otodus obliquus Agassiz; Phosphates Formation; Early Eocene; Morocco.

unserrated cutting edge

Upper tooth

Lower tooth

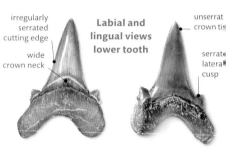

irregularly serrated cutting edge

wide crown neck

Labial and lingual views lower tooth

unserrat crown ti

serrate lateral cusp

Otodus aksuaticus (Menner); Aktulagay Formation; Early Eocene; Kazakhstan.

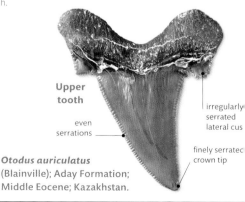

Upper tooth

even serrations

irregularly serrated lateral cus

finely serrated crown tip

Otodus auriculatus (Blainville); Aday Formation; Middle Eocene; Kazakhstan.

Typical length
Eocene 16 ft (5 m)
Pliocene 52 ft (16 m)

Range: Paleocene–Pliocene	Distribution: Worldwide	Occurrence:

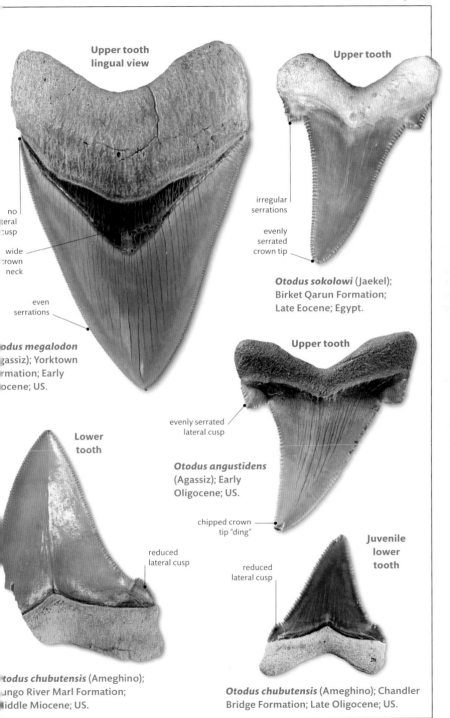

Upper tooth lingual view

no lateral cusp

wide crown neck

even serrations

Otodus megalodon (Agassiz); Yorktown Formation; Early Pliocene; US.

Upper tooth

irregular serrations

evenly serrated crown tip

Otodus sokolowi (Jaekel); Birket Qarun Formation; Late Eocene; Egypt.

Upper tooth

evenly serrated lateral cusp

Otodus angustidens (Agassiz); Early Oligocene; US.

chipped crown tip "ding"

Lower tooth

reduced lateral cusp

Otodus chubutensis (Ameghino); Pungo River Marl Formation; Middle Miocene; US.

Juvenile lower tooth

reduced lateral cusp

Otodus chubutensis (Ameghino); Chandler Bridge Formation; Late Oligocene; US.

| Group: CARCHARINIFORMES | Subgroup: CARCHARINIDAE | Informal name: Tiger shark |

Galeocerdo

The tooth of this shark consists of a crown with a finely serrated cusp and more coarsely serrated blade. The root bears a shallow groove and several vascular openings (foraminae).

HABITAT This genus inhabits coastal waters.

Galeocerdo cuvier
(Peron & LeSueur);
Yorktown Formation;
Early Pliocene; US.

Typical length 16½ ft (5 m)

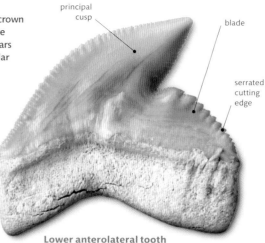

principal cusp

blade

serrated cutting edge

Lower anterolateral tooth

| Range: Eocene–Recent | Distribution: Worldwide | Occurrence: |

| Group: CARCHARINIFORMES | Subgroup: HEMIGALEIDAE | Informal name: Snaggletooth shark |

Hemipristis

The upper teeth of this shark are triangular, with serrated cutting edges. The symphyseal tooth (closest to the lower jaw junction) is almost symmetrical, while side teeth become increasingly inclined away from the center. Lower teeth are slimmer, with V-shaped roots. The symphyseal tooth has few serrations, but lateral teeth become progressively serrated and broad.

HABITAT *Hemipristis* lived in warm coastal waters.
REMARK Many fossil sharks, like *Hemipristis*, have differing upper and lower teeth.

Hemipristis serra Agassiz;
Pungo River Formation;
Middle Miocene; US.

Upper teeth

symphyseal tooth

Reconstructed dentition

two antero-lateral teeth

posterior tooth

Lower teeth

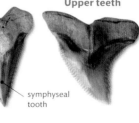

Typical length 16½ ft (5 m)

symphyseal tooth

three antero-lateral teeth

posterior tooth

| Range: Eocene–Recent | Distribution: Worldwide | Occurrence: |

| Group: SCLERORHYNCHIFORMES | Subgroup: SCLERORHYNCHIDAE | Informal name: Saw shark |

Ischyrhiza

Now extinct, *Ischyrhiza* was a genus of saw shark with small oral teeth and large teeth on the rostrum (snout). The cap (crown) of the rostral teeth is extended, flattened, and pointed, with cutting edges front and back. At the base of the posterior cutting edge is a small bulge. The root has a folded upper and lower surface and a divided base.

HABITAT An inhabitant of inshore waters, this genus could tolerate a range of salinities.
REMARK *Ischyrhiza* used its toothed rostrum to comb food—probably worms and crustaceans—from the sediment, and possibly for defense.

Ischyrhiza nigeriensis (Tabaste); Dukamaje Formation; Late Cretaceous; Niger.

Typical length 7¼ft (2.2 m)

cap

surface polished by wear

posterior cutting edge

anterior cutting edge

bulge at cap base

folded root

Ventral view   **Posterior view**

| Range: Late Cretaceous | Distribution: Americas, Africa, Europe | Occurrence: 🐚🐚🐚🐚 |

| Group: MYLIOBATIFORMES | Subgroup: MYLIOBATIDAE | Informal name: Eagle ray |

Myliobatis

The teeth of the eagle ray, *Myliobatis*, are arranged in an upper and lower plate, each with seven files. The middle file is wide and hexagonal, with a smooth, slightly convex oral surface, and crinkled surfaces on the lips and tongues. The three side files are narrow and either hexagonal, pentagonal, or triangular.

HABITAT *Myliobatis* inhabits warm, shallow marine environments, living on crustaceans, mollusks, and small fish.
REMARK After death, the plates usually disintegrate.

Myliobatis toliapicus Agassiz; London Clay; Early Eocene; UK.

Lower tooth plate

Typical length 5 ft (1.5 m)

small side file

large middle file

| Range: Paleocene–Recent | Distribution: Worldwide | Occurrence: 🐚🐚🐚🐚🐚 |

Group: MYLIOBATIFORMES	Subgroup: DASYATIDAE	Informal name: Stingray

Heliobatis

Heliobatis was a stingray with a rounded disk; pointed snout; and long, barbed tail. A series of hooked, dermal denticles ran along the dorsal midline. The whiplike tail was armed with up to three barbed spines. The 90 pectoral-fin rays almost met in front of the skull, with the 16 pelvic-fin rays completing the circle behind. The teeth were small and tetrahedral, with a relatively flat occlusal surface. Males had a pair of pelvic claspers. The tail consisted of between 170 and 190 fully calcified vertebrae.

HABITAT Living in freshwater streams and lakes, this stingray probably fed on crayfish, prawns, and other invertebrates.

REMARK When not feeding, *Heliobatis* would lie partially buried in soft sediment. The poisonous tail spines could inflict serious injury to any potential predator.

Heliobatis radians
Marsh; Green
River Formation;
Early Eocene; US.

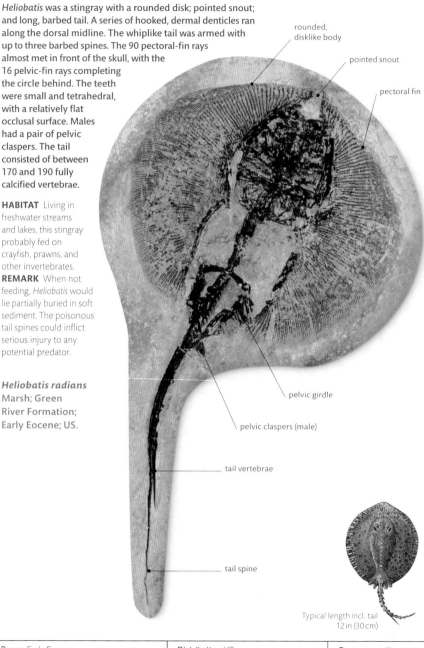

rounded, disklike body

pointed snout

pectoral fin

pelvic girdle

pelvic claspers (male)

tail vertebrae

tail spine

Typical length incl. tail
12 in (30 cm)

Range: Early Eocene	Distribution: US	Occurrence:

Group: COCHLIDONTIFORMES	Subgroup: COCHLIODONTIDAE	Informal name: Rabbitfish

Sandalodus

Although known only from isolated tooth plates, this fish is believed to have reached up to 6½ft (2 m) in length. The lower teeth are curved strongly inward; the upper teeth are extended triangles in outline, with pointed anterior ends and broad, rounded posterior ends. The biting surface is traversed by longitudinal ridges, giving the upper surface of the plate an undulating cross-section. Vertical pillars of hard dentine rise from the tooth plate surface.

HABITAT *Sandalodus* used its teeth to crush thick-shelled invertebrates—and possibly corals—in warm, shallow, shelf seas.

Sandalodus morrisii Davis; Carboniferous Limestone; Early Carboniferous; UK.

Upper tooth plate

grinding surface

posterior

Typical length 6½ft (2 m)

Range: Carboniferous	Distribution: Europe	Occurrence:

Group: CHIMAERIFORMES	Subgroup: CALLORHINCHIDAE	Informal name: Rabbitfish

Edaphodon

This rabbitfish is known only from isolated tooth plates. The dentition consisted of one pair of lower tooth plates and two pairs of upper tooth plates. The crushing surface of each bony plate is made up of localized areas of specialized dentine (tritors), consisting of hard dentine pillars positioned normally to the tooth plate surface. These were effective in dealing with thick-shelled marine invertebrates.

anterior tip

Lower tooth plate

middle of jaw

HABITAT Modern rabbitfish inhabit shallow to deep, cool marine waters.

Edaphodon bucklandi Agassiz; Bracklesham Beds; Middle Eocene; UK.

central specialized crushing area

Typical length 3½ft (1.1 m)

Range: Cretaceous–Pliocene	Distribution: Worldwide	Occurrence:

ACANTHODIANS

THESE FISHES WERE among the earliest known gnathostomes and flourished in the Devonian. They varied in size from 4 in (10 cm) to over 6½ ft (2 m) and were active swimmers, with long, tapered bodies protected by scales. A strongly heterocercal tail provided considerable thrust when swimming, and all paired and median fins (except the tail) had strong, immovable spines on their leading edges. The gills were covered by a large scale (operculum) and the teeth often developed in whorls.

Group: ACANTHODEI	Subgroup: DIPLACANTHIDAE	Informal name: Spiny shark

Diplacanthus

The moderately deep body of this genus narrows at the back into a powerful tail. The two dorsal fins and pectoral fins are supported by spines. Two pairs of intermediate spines lie along the underside of the body.

HABITAT *Diplacanthus* lived in shallow lakes.

Typical length
4 in (10 cm)

Diplacanthus
sp.; Lower Old
Red Sandstone;
Devonian; UK.

pectoral-fin
spine

dorsal spine

tail

Range: Devonian	Distribution: Europe	Occurrence: ⌇⌇⌇ ⌇⌇⌇

Group: ACANTHODEI	Subgroup: ACANTHOESSIDAE	Informal name: Spiny shark

Cheiracanthus

This moderate-sized acanthodian has a robust and fairly deep body, with large eyes placed well forward on the sides, and a heterocercal tail. It has single dorsal and anal fins and paired pectoral and pelvic fins, all protected by spines. Paired intermediate spines were lacking. The scales were small and ornamented.

HABITAT With its gaping jaws, this was an efficient midlevel and surface filter feeder in fresh water.

scale-covered
body

Cheiracanthus **sp.;** Middle
Old Red Sandstone;
Devonian; UK.

nodule

Typical cup
diameter
12 in (30 cm)

tail

ventral
intermediate spines

jaw

Range: Devonian	Distribution: Europe, Antarctica	Occurrence: ⌇⌇⌇

OSTEICHTHYANS

FISH BELONGING TO this class are characterized by a bony internal skeleton. Two subgroups can be identified by their fin structure: the actinopterygians had fins supported by bony rods (radials), while the sarcopterygians had fleshy fins, which were supported by a single bone at the base. Actinopterygians (Devonian to Recent) are very diverse and include the majority of present-day marine and freshwater fishes. The sarcopterygians dominated during the Devonian and include the lungfish and coelacanths.

Group: CERATODONTIFORMES	Subgroup: CERATODONTIDAE	Informal name: Lungfish

Ceratodus

A fossil relative of the extant Australian lungfish, *Ceratodus* had two upper and two lower tooth plates, anchored to the jaws. Up to six prominent, diagonal ridges with deep, intervening troughs cross the plates, from the inner posterior angle to the outer margin.

HABITAT *Ceratodus* lived in fresh water, feeding on the bottom.

ridge

Upper plates

meeting point of lower jaws

bony support

Lower plates

margin

Typical length 20 in (50 cm)

Ceratodus tiguidensis Tabaste; Ihrazer Shale; Middle Jurassic; Republic of Niger.

Range: Early Triassic–Paleocene	Distribution: Worldwide	Occurrence:

Group: PALEONISCIFORMES	Subgroup: BIRGERIIDAE	Informal name: Paleoniscid

Birgeria

Both jaws of this fish carried a row of large, predatory fangs flanked by smaller teeth. All were upright and conical, with a translucent enamel cap, folded root wall, and striated base.

HABITAT *Birgeria* was a formidable marine predator.

outer surface covered by round projections

Birgeria acuminata (Agassiz); Late Triassic; UK.

Upper jaw

smaller, intervening teeth

fang

Typical length 20 in (50 cm)

Range: Triassic	Distribution: Europe, Greenland	Occurrence:

| Group: PALAEONISCIFORMES | Subgroup: PALAEONISCIDAE | Informal name: Palaeoniscid |

Palaeoniscus

A long fish with a spindle-shaped body, *Palaeoniscus* is covered by a coat of overlapping scales. The tail is strongly heterocercal, and the dorsal edge was marked by a strengthening ridge of scales, which acted as a cutwater.

Palaeoniscus magnus
Agassiz; Kupferschiefer;
Late Permian; Germany.

base of tail

HABITAT This strong-swimming fish was a marine dweller.

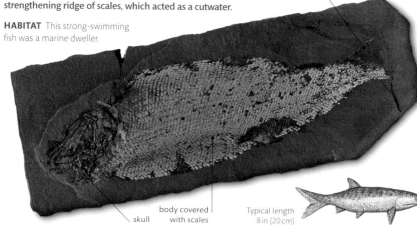

skull

body covered with scales

Typical length
8 in (20 cm)

| Range: Permian–Triassic | Distribution: Worldwide | Occurrence: |

| Group: SEMIONOTIFORMES | Subgroup: SEMIONOTIDAE | Informal name: Neopterygian |

Lepidotes

This fossil fish is rarely found as a whole fish; usually, just isolated scales and bony plates are preserved. The skull bones were ornamented with knoblike projections; a massive operculum covered the gills; and the mouth was armed with hemispherical, crushing teeth. The moderately deep body was protected by shiny, thickly enameled scales, arranged in longitudinal rows.

Lepidotes sp.; Oxford
Clay; Late Jurassic; UK.

rows of
rhomboidal
scales

HABITAT *Lepidotes* fed on thick-shelled, bottom-dwelling invertebrates in shallow marine seas, lagoons, and freshwater lakes.

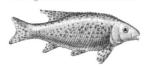

Typical length 5½ ft (1.7 m)

Fragment of flank

enamel-covered
outer surface

| Range: Triassic–Cretaceous | Distribution: Worldwide | Occurrence: |

Group: PYCNODONTIFORMES	Subgroup: PYCNODONTIDAE	Informal name: Pycnodont

Gyrodus

This was a very deep-bodied fish with an almost circular outline. It had symmetrically placed dorsal and anal fins; a deeply notched, symmetrical tail fin; and reduced pelvic fins placed well forward. A delicate network of fine bones supported the body internally, and deep, rectangular scales covered the body surface.

HABITAT *Gyrodus*'s dense pavement of rounded teeth suggests it may have browsed on a diet of coral or other hard-bodied prey.

Lower jaw

densely packed teeth

longitudinal tooth rows

Gyrodus cuvieri
Agassiz;
Kimmeridge Clay;
Late Jurassic; UK.

outer margin

Typical length 1 ft (30 cm)

Range: Late Jurassic–Early Cretaceous	Distribution: Worldwide	Occurrence:

Group: SEMIONOTIFORMES	Subgroup: DAPEDIIDAE	Informal name: Neopterygian

Dapedium

This deep-bodied fish was oval in outline and laterally compressed. Its body had larger dorsal, anal, and pelvic fins than *Gyrodus* and a fan-shaped tail. The skull bones are distinctively ornamented with round projections, and the eyes were surrounded by a ring of plates. At the front of the mouth are nipping, upright teeth, which give way to a pavement of circular crushing teeth at the back.

Dapedium politum
Agassiz; Lower Lias;
Early Jurassic; UK.

skull roof

eye socket

HABITAT Fishes of this genus probably fed on a variety of invertebrates, such as crustaceans, in warm, shallow seas.

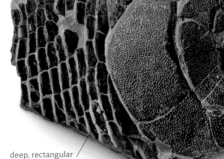

deep, rectangular body scales

Partial skull

Typical length 16 in (40 cm)

operculum over gills

cheek region

Range: Late Triassic–Jurassic	Distribution: Europe, Asia	Occurrence:

Group: PACHYCORMIFORMES	Subgroup: PACHYCORMIDAE	Informal name: Neopterygian

Pachycormus

This large fish had a powerfully built body and deeply notched, homocercal tail. The pelvic fins were much reduced, while the scythelike pectoral fins were greatly elongated.

Pachycormus macropterus (Blainville); Early Jurassic; France.

HABITAT This fast, open-water predator inhabited shallow marine waters.

REMARK This specimen is preserved three-dimensionally in a calcareous nodule.

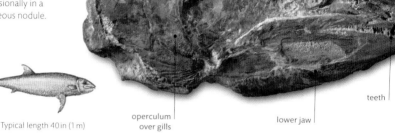

skull-roof bones

teeth

operculum over gills

lower jaw

Typical length 40 in (1 m)

Range: Jurassic	Distribution: Europe	Occurrence:

Group: SILURIFORMES	Subgroup: BAGRIDAE	Informal name: Catfish

Sperata

This fossil catfish is only known from its skull. It is likely that it had a fairly long body with a forked tail, a moderately long anal fin well separated from the tail, and a front dorsal fin supported by a long, spiny ray. A second dorsal fin (the adipose fin) consisted of an unsupported flap of fatty tissue. The body had no scales. The palate and both jaws carried teeth. Elongate, tactile sense organs called barbels originated on the upper jaw.

Sperata aor (Bucanon); Early Pliocene; India.

skull roof

Skull

HABITAT This bottom-dwelling fish preyed on invertebrates.

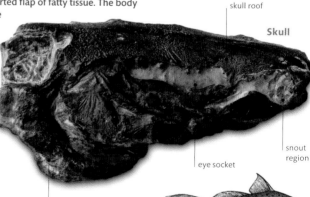

snout region

eye socket

cheek and gill covering bones

Estimated length 20 in (50 cm)

Range: Pliocene–Recent	Distribution: India	Occurrence:

| Group: ASPIDORHYNCHIFORMES | Subgroup: ASPIDORHYNCHIDAE | Informal name: Neopterygian |

Aspidorhynchus

This long, scale-covered fish was driven by a notched, homocercal tail and separated pelvic and pectoral fins. Its triangular, pointed skull has a well-developed snout, forward-placed eyes, and sharp teeth.

HABITAT This genus inhabited shallow, subtropical seas.

Aspidorhynchus acutirostris
Blainville; Lithographic Limestone; Late Jurassic; Germany.

Anterior trunk

eye socket

pointed skull

rostrum

pectoral fin

Typical length 20 in (50 cm)

| Range: Late Jurassic–Early Cretaceous | Distribution: Europe | Occurrence: |

| Group: SALMONIFORMES | Subgroup: ENCHODONTIDAE | Informal name: Saber-toothed herring |

Enchodus

The dorsal midline of the body carried large, bony plates. A short dorsal fin was followed by a small, adipose (fatty) fin, and large, bony plates appeared along the dorsal midline. The lower jaw carries a row of large teeth, flanked by a smaller tooth row, with a large front fang on the upper jaw.

HABITAT *Enchodus* was a marine predator.
REMARK Isolated teeth and jaw fragments of *Enchodus* are common in the Late Cretaceous chalks of Kansas and Alabama.

Enchodus lewesiensis (Mantell); Lower Chalk; Late Cretaceous; UK.

Skull

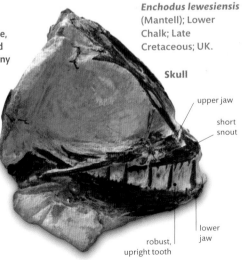

upper jaw

short snout

lower jaw

robust, upright tooth

Typical length 20 in (50 cm)

| Range: Late Cretaceous–Eocene | Distribution: Worldwide | Occurrence: |

Group: LABRIFORMES	Subgroup: LABRIDAE	Informal name: Elopiform

Phyllodus

This fish is known primarily from its convex tooth plates. The oval central teeth are the largest and are organized into a rough, longitudinal row in the middle. These are flanked by long, slightly smaller crowns, giving way at the sides to almost concentric rows of smaller, near-circular teeth and occasional ovoid teeth.

HABITAT *Phyllodus* probably fed on thick-shelled, bottom-dwelling marine invertebrates, such as mollusks. As the fossil fish is also found in lagoonal and brackish deposits, it is believed that it may have been able to tolerate salinity.
REMARK The affinities of the genus to other genera are uncertain.

Upper pharyngeal tooth plate

small marginal teeth

laterally expanded central teeth

Phyllodus toliapicus Agassiz; Blackheath Beds; Early Eocene; UK.

Estimated length 16 in (40 cm)

Range: Eocene	Distribution: N. America, Europe	Occurrence: ⫯⫯⫯

Group: ICHTHYODECTIFORMES	Subgroup: UNCLASSIFIED	Informal name: Tarpan

Pachythrissops

Pachythrissops was a spindle-shaped fish, with a short skull and a prominent symmetrical, forked tail. The single dorsal and anal fins are much larger than the paired pectoral and pelvic fins. Thin, rounded scales protected the body.

Pachythrissops furcatus (Eastman); Solnhofen Limestone; Late Jurassic; Germany.

HABITAT *Pachythrissops* used its small, conical teeth to feed on small marine mollusks, arthropods, and small fish.

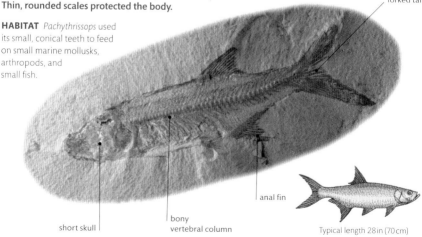

forked tail

anal fin

short skull

bony vertebral column

Typical length 28 in (70 cm)

Range: Late Jurassic–Early Cretaceous	Distribution: Europe	Occurrence: ⫯⫯⫯

| Group: ELOPIFORMES | Subgroup: ELOPIDAE | Informal name: Elopiform |

Rhacolepis

The body is spindle-shaped, covered with small rhomboidal scales. The skull is pointed. The lower jaw is long and upwardly inclined, with a single row of small teeth.

Rhacolepis buccalis
Agassiz; Santana Formation;
Early Cretaceous; Brazil.

HABITAT *Rhacolepis* lived in shallow marine conditions. It is commonly found in calcareous (limestone) nodules, preserved in three dimensions.

tail

skull and opercular bones

Typical length 8 in (20 cm)

angled lower jaw

scales

| Range: Early Cretaceous | Distribution: S. America | Occurrence: |

| Group: ELLIMMICHTHYIFORMES | Subgroup: ELLIMMICHTHYIDAE | Informal name: Herring |

Diplomystus

Diplomystus has a moderately deep body, a homocercal tail, single dorsal and anal fins, and a pelvic fin situated directly beneath the dorsal fin. The scales are thin and ovoid. The strongly upturned mouth is typical of a surface-water feeder.

HABITAT *Diplomystus* was a common inhabitant of some North American Eocene lakes.

REMARK Some specimens have smaller fishes preserved in their mouths or intestines.

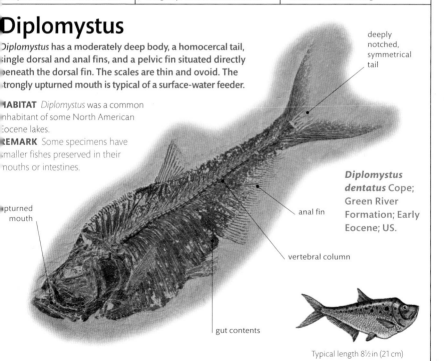

deeply notched, symmetrical tail

Diplomystus dentatus Cope; Green River Formation; Early Eocene; US.

upturned mouth

anal fin

vertebral column

gut contents

Typical length 8½ in (21 cm)

| Range: Cretaceous–Eocene | Distribution: N. and S. America | Occurrence: |

| Group: CYPRINIFORMES | Subgroup: CYPRINIDAE | Informal name: Dace |

Leuciscus

This dace has a small but extended body, small fins, relatively large scales, and a forked tail. It is a specialized herbivore with toothless jaws. The front vertebrae and ribs (known as the Weberian ossicles) are movable and, by transmitting vibrations from the swim bladder to the inner ear, improve sensitivity to high-frequency sound. When injured, modern species of *Leuciscus* release a chemical from epidermal alarm cells, eliciting a fright reaction in related fish, causing them to scatter and swim to the bottom.

HABITAT *Leuciscus* inhabits freshwater streams and lakes.
REMARKS This specimen shows a mass mortality, with subsequent irregular alignment of the bodies by the action of currents.

Leuciscus pachecoi
Gomez; Miocene;
Spain.

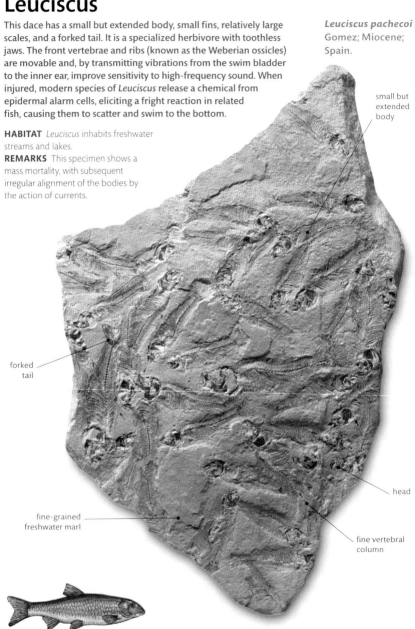

small but extended body

forked tail

head

fine-grained freshwater marl

fine vertebral column

Typical length 3½ in (9 cm)

| Range: Oligocene–Recent | Distribution: N. America, Asia, Africa | Occurrence: |

Group: BERYCIFORMES	Subgroup: BERYCIDAE	Informal name: Alfonsino

Centroberyx

Some fossil fish, like this one, may be identified by their otoliths, which are concentrically laminated, aragonitic structures located in the inner ear. Otoliths are embedded in ciliated sense organs. They detect changes in body position. Of the three pairs, the sacculith (shown) is the largest.

HABITAT *Centroberyx* lived in shoals in moderately deep oceans.

dorsal rim

sulcus

Otolith

typical length 24 in (60 cm)

typical orange color

Centroberyx eocenicus (Frost); London Clay; Early Eocene; UK.

Range: Late Cretaceous–Oligocene	Distribution: Worldwide	Occurrence:

Group: BERYCIFORMES	Subgroup: TRACHICHTHYIDAE	Informal name: Slimehead

Hoplopteryx

Hoplopteryx has a dorsal fin supported by nine unjointed, bony fin rays; a deeply forked, homocercal tail; a moderately developed anal fin; and a pelvic fin located well forward. The snout is quite short, the eyes are fairly large, and both jaws of the upturned mouth hold small teeth.

HABITAT This was a marine fish, living in shallow chalk seas.

Hoplopteryx lewesiensis (Mantell); Lower Chalk; Late Cretaceous; UK.

fin rays

vertebral column

eye socket

bones covering gills

small pectoral fin

Typical length 10½ in (27 cm)

Range: Late Cretaceous	Distribution: Northern hemisphere	Occurrence:

Group: PERCIFORMES	Subgroup: MORONIDAE	Informal name: Temperate bass

Cockerellites

This fossil perch has a deep, oval body protected by dorsal and anal spines, and a fan-shaped tail. It has a slightly upturned, protruding lower jaw. The two jaws, as well as the enlarged bones in the mouth, are covered with fine teeth.

HABITAT *Cockerellites* lived in freshwater streams and lakes, feeding on snails and crustaceans.

Cockerellites liops
Cope; Green River Formation; Middle Eocene; US.

vertebrae   stout dorsal spines   eye socket

unforked tail fin

Typical length
6 in (15 cm)

lower jaw

Range: Eocene	Distribution: N. America	Occurrence:

Group: PERCIFORMES	Subgroup: SCOMBRIDAE	Informal name: Mackerel

Wetherellus

Although known from the cranial skeleton only, this marine perch is thought to have been around 10 in (25 cm) long. Its moderately large eye, containing sclerotic plates, is centrally placed in the shallow skull. The jaws are armed with a row of upright, pointed teeth flanked by a smaller, marginal series.

HABITAT *Wetherellus* lived in moderately deep oceans.
REMARK This skull is preserved in a phosphatic nodule.

Wetherellus cristatus Casier; London Clay; Early Eocene; UK.

eye socket

sclerotic ring

upper jaw

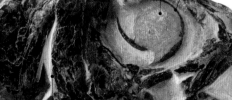

Estimated length
10 in (25 cm)

operculum over gill

gaping lower jaw

Range: Early Eocene	Distribution: Europe	Occurrence:

TETRAPODS

AMPHIBIANS WERE THE first tetrapods to colonize the land, over 400 million years ago. Although they were able to live on land, their eggs were laid in water, hence they were only semiterrestrial. From the late Carboniferous period onward, the evolution of tetrapods followed two major pathways. One led to the lissamphians, frogs, salamanders, and caecilians. The second pathway gave rise to the amniotes, which developed a waterproof egg, allowing them to breed on land and so become completely terrestrial, unlike today's semiterrestrial amphibians.

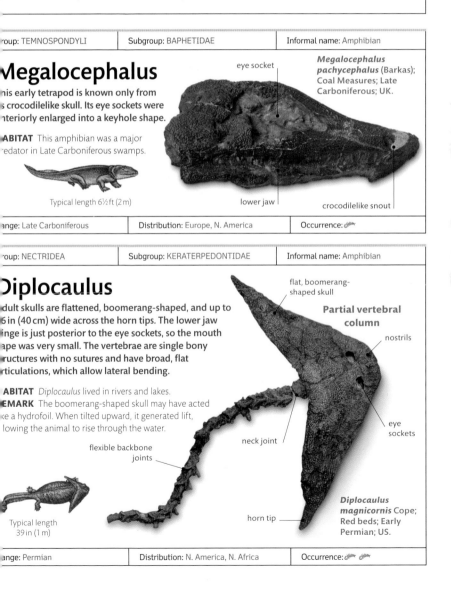

Group: TEMNOSPONDYLI	Subgroup: BAPHETIDAE	Informal name: Amphibian

Megalocephalus

This early tetrapod is known only from its crocodilelike skull. Its eye sockets were anteriorly enlarged into a keyhole shape.

HABITAT This amphibian was a major predator in Late Carboniferous swamps.

Typical length 6½ft (2 m)

eye socket

Megalocephalus pachycephalus (Barkas); Coal Measures; Late Carboniferous; UK.

lower jaw

crocodilelike snout

Range: Late Carboniferous	Distribution: Europe, N. America	Occurrence:

Group: NECTRIDEA	Subgroup: KERATERPEDONTIDAE	Informal name: Amphibian

Diplocaulus

Adult skulls are flattened, boomerang-shaped, and up to 16 in (40 cm) wide across the horn tips. The lower jaw hinge is just posterior to the eye sockets, so the mouth gape was very small. The vertebrae are single bony structures with no sutures and have broad, flat articulations, which allow lateral bending.

HABITAT *Diplocaulus* lived in rivers and lakes.
REMARK The boomerang-shaped skull may have acted like a hydrofoil. When tilted upward, it generated lift, allowing the animal to rise through the water.

flexible backbone joints

Typical length 39 in (1 m)

flat, boomerang-shaped skull

Partial vertebral column

nostrils

neck joint

eye sockets

horn tip

Diplocaulus magnicornis Cope; Red beds; Early Permian; US.

Range: Permian	Distribution: N. America, N. Africa	Occurrence:

| Group: TEMNOSPONDYLI | Subgroup: BRANCHIOSAURIDAE | Informal name: Amphibian |

Apateon

This small, neotenous amphibian resembles a salamander. Its skull usually measures between ⁵⁄₁₆ and 1 in (8 and 24mm) in length. The posterior region of the skull behind the eyes is very short. There were three pairs of long, feathery external gills. The ribs were very reduced and the bones of the wrists and ankles are unossified. The hand had four fingers. The tail bore a long, deep fin.

HABITAT *Apateon* was a totally aquatic amphibian, inhabiting semipermanent lakes and ponds.
REMARK Fluctuations in seasonal conditions caused mass deaths of thousands of individuals. These are now preserved in fine-grained, freshwater limestones.

Apateon pedestris Meyer; Freshwater limestone; Early Permian; Germany

short skull

feathery external gills

body outline

tail fin

Typical length 4¾ in (12 cm)

| Range: Early Permian | Distribution: Europe | Occurrence: |

| Group: TEMNOSPONDYLI | Subgroup: TREMATOSAURIDAE | Informal name: Amphibian |

Aphaneramma

The narrow skull has an extremely elongate, slender snout. The long, slitlike nostrils are situated halfway between the snout tip and the eye sockets. The eye sockets are round and have slightly raised rims. The lateral line canals are very prominent, deep channels; the canals running above and below the eye sockets do not join together. The skull is ornamented with long ridges of bone.

HABITAT *Aphaneramma* was a specialist fish-eater, living in fresh- to brackish water habitats.

rounded snout tip

long, narrow snout

Skull

nostril

lateral line canal

eye socket

striated bone ornament

Aphaneramma sp.; Red beds; Early Triassic; US.

Typical length 6½ ft (2 m)

| Range: Early Triassic | Distribution: Worldwide | Occurrence: |

| Group: TEMNOSPONDYLI | Subgroup: BRACHYOPIDAE | Informal name: Amphibian |

Batrachosuchus

The broad, triangular, short-faced skull of *Batrachosuchus* has nostrils set close together. The eye sockets are widely separated, the upper jaw bone forming part of the outer rim.

HABITAT *Batrachosuchus* was a neotenous surface swimmer that probably fed on small prey by suction gulping.

Typical length
20 in (50 cm)

Batrachosuchus watsoni Haughton; Karroo Beds; Early Triassic; South Africa.

nostrils

eye sockets
far forward

lateral
line

Skull

pit
and ridge
ornament

| Range: Triassic | Distribution: Africa | Occurrence: |

| Group: URODELA | Subgroup: CRYPTOBRANCHIDAE | Informal name: Giant salamander |

Andrias

This is the largest salamander, up to 7½ ft (2.3 m) long. The skull is short, broad, and massive in adults. There are four fingers, five toes, and a short tail.

HABITAT Modern species of *Andrias* live in rivers and large streams, never leaving the water; fossil species probably lived the same way.

Andrias scheuchzeri (Holl); Late Miocene; Germany.

short, broad skull

tail

disarticulated foot

hind limb

straight ribs

four-fingered hand

Typical length
6½ ft (2 m)

| Range: Miocene–Recent | Distribution: Europe, Japan, N. America | Occurrence: |

Group: EMBOLOMERA	Subgroup: ARCHERIIDAE	Informal name: Amphibian

Archeria

These slender, long-bodied swimmers grew up to about 6½ft (2 m) long. The skull is high posteriorly, with a small horn above a deep cheek notch. The snout is depressed and spatulate. At least 40 small chisel-shaped teeth on each side of the jaw gave a long biting edge like a hacksaw blade.

Archeria crassidisca (Cope); Red beds; Early Permian; US.

HABITAT *Archeria* lived in lakes and fed on soft-bodied invertebrates.

spatulate snout

large teeth at snout tip

eye socket

cheek notch

small, chisel-like teeth

Typical length 6½ft (2 m)

Range: Early Permian	Distribution: N. America	Occurrence:

LISSAMPHIBIANS

LISSAMPHIBIANS are a group of smooth-skinned, non-amniote tetrapods that include the frogs and toads, salamanders and newts, the limbless caecilians as well as the extinct salamander-like albanerpetontids.

Group: ANURA	Subgroup: RANIDAE	Informal name: Frog

Rana

These true frogs have slender, streamlined bodies and a pointed head with large eyes and eardrums. The skeleton is extensively modified for swimming and jumping. The body is short and inflexible; the shoulder girdle acts as a shock absorber; and the long, powerful hind limbs and large feet provide thrust in water or from the ground.

HABITAT Most species of the genus lay eggs in water and have free-swimming tadpoles. Adults are mainly aquatic, inhabiting regions ranging from the rain forest to temperate, high-latitude habitats.

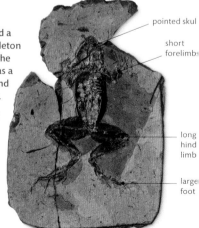

pointed skul

short forelimb

long hind limb

large foot

Rana pueyoi Navas; Red shales; Oligocene; Spain.

Typical length 3½ in (9 cm)

Range: Eocene–Recent	Distribution: Worldwide, except Australasia	Occurrence:

AMNIOTES

AMNIOTES are a successful and diverse group of tetrapods. They share the possession of an amniotic egg, which freed them from a dependence on water for reproduction. There are two major amniote lineages: the Synapsids (mammals and their extinct close relatives) and the Reptiles (including the dinousaurs and birds).

REPTILES

THE CLASS REPTILA is divided into three subclasses (the Anapsida, Diapsida, and Synapsida) according to the number of openings in the skull roof and the configuration of the skull bones behind the eye sockets. The early forms were generally small lizard-like creatures. However, they gave rise to the dinosaurs and birds.

Group: PROCOLOPHONIA	Subgroup: PROCOLOPHONIDAE	Informal name: Procolophonid

Procolophon

Evolving from the early members of the anapsid line, the lizardlike, triangular-headed procolophonids retained many primitive features in their skeletons. For example, the skull has a pineal foramen for the pineal or "third" eye that senses only light. This is more in keeping with the anapsids' amphibious ancestors, while the jaws and palate have numerous rows of simple teeth. The heavy limbs and relatively massive fore- and hindlimb girdles indicate slow movement and a sprawling gait.

HABITAT All appear to have been land-living vegetarians.
REMARK The family first appeared in the Early Triassic but failed to survive beyond it.

Procolophon trigoniceps Owen; Karoo Formation; Early Triassic; South Africa.

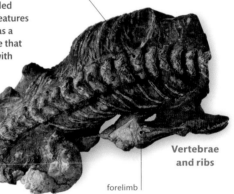

backbone

Vertebrae and ribs

forelimb

Skull

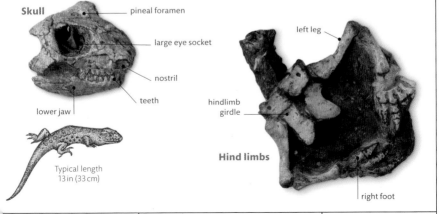

pineal foramen

large eye socket

nostril

teeth

lower jaw

left leg

hindlimb girdle

Hind limbs

right foot

Typical length 13 in (33 cm)

Range: Early Triassic	Distribution: S. Africa	Occurrence:

| Group: PROCOLOPHONIA | Subgroup: PAREIASAURIDAE | Informal name: Pareiasaurid |

Elginia

This small example of the pareiasaur family is distinguished by the elaborate bony processes produced from the triangular-shaped skull, which probably acted either as a form of defense against would-be predators or as a display mechanism while courting.

HABITAT The four-footed *Elginia* lived on land plants. It became extinct during the Late Permian.

hornlike processes

Skull

pineal foramen

eye sockets

nostril

Typical length 5 ft (1.5 m)

Elginia mirabilis
Newton; Elgin Sandstone;
Late Permian; UK.

| Range: Late Permian | Distribution: Europe | Occurrence: |

| Group: SAUROPSIDA | Subgroup: MESOSAURIDAE | Informal name: Mesosaurid |

Stereosternum

The aquatic preference of this anapsid is evident in the long, needlelike teeth; the elongated head and body, which contain 10 neck vertebrae; and the short limbs and flattened tail.

HABITAT It has been suggested that the long, fine teeth of this small carnivore combined to act as a sieve, enabling it to filter out the small, swimming crustaceans abundant in its marine environment.

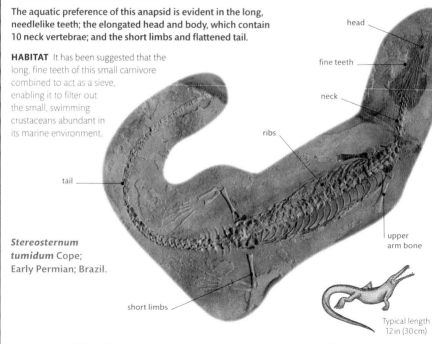

head

fine teeth

neck

ribs

tail

upper arm bone

Stereosternum tumidum Cope;
Early Permian; Brazil.

short limbs

Typical length 12 in (30 cm)

| Range: Early Permian | Distribution: S. Africa, S. America | Occurrence: |

| Group: SPHENODONTA | Subgroup: SPHENODONTIDAE | Informal name: Sphenodontid |

Homeosaurus

A small, lizardlike reptile, *Homeosaurus* resembles the last surviving member of its order—the tuatara of New Zealand—in many details. The group to which it belongs differs from the lizards by having skulls with a rigid articulation bone (quadrate). Also characteristic is the attachment of the few marginal teeth, which are firmly fused to the edge of the jaws. These were not replaced when worn, but were added at the rear as growth continued. Further teeth are present in the palate, and the snout is turned down to produce a chisel-like cutting edge.

HABITAT *Homeosauros* was probably a land dweller, living on plants and insects.

REMARK First appearing during the Triassic, the order became widespread during the early part of the Mesozoic, before declining to the single species still alive today.

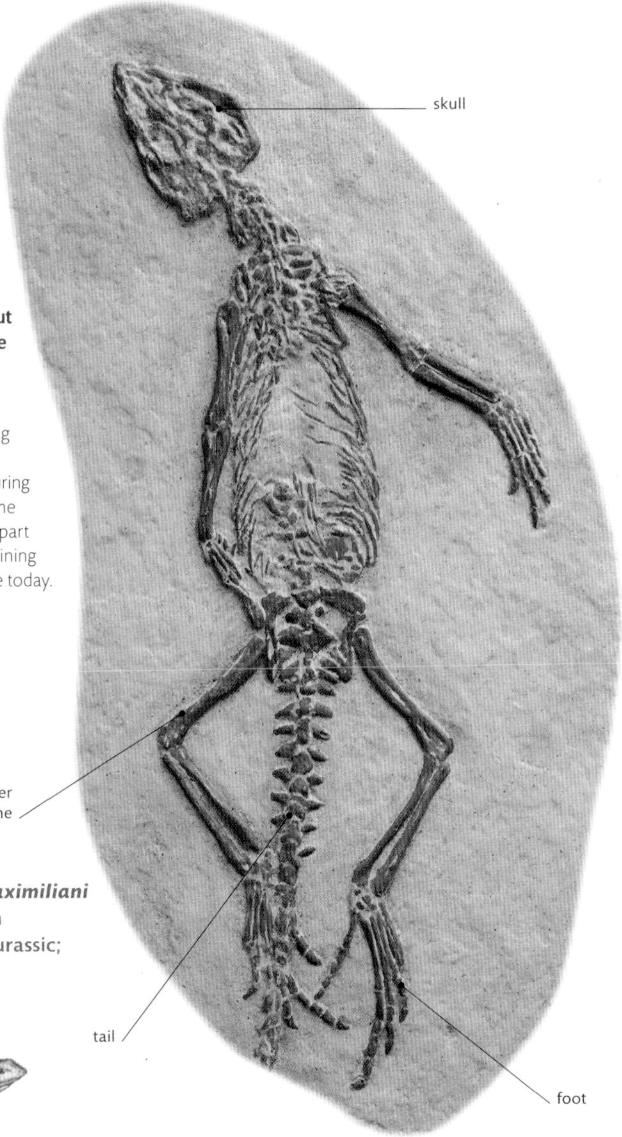

skull

upper leg bone

Homeosaurus maximiliani Meyer; Solnhofen Limestone; Late Jurassic; Germany.

tail

foot

Typical length 8 in (20 cm)

| Range: Late Jurassic | Distribution: Europe | Occurrence: |

| Group: EOSUCHIA | Subgroup: KUEHNEOSAURIDAE | Informal name: Flying lizard |

Kuehneosuchus

This primitive, gliding, lizardlike diapsid had very long, fixed ribs, giving a "wingspan" of about 12 in (30 cm). It is thought that these ribs were attached to each other by tissue, thus forming a continuous membrane, which would have acted as a gliding surface reminiscent of present-day "flying lizards" such as *Draco*.

HABITAT *Kuehneosuchus* is thought to have been insectivorous and lived in open woodland.

Kuehneosuchus latissimus
Robinson; Fissure fill; Late Triassic; UK.

three vertebrae

long, hollow rib

rib

upper thigh bone

fragments

Typical "wingspan"
12 in (30 cm)

| Range: Late Triassic | Distribution: Europe | Occurrence: |

| Group: SQUAMATA | Subgroup: AMPHISBAENIDAE | Informal name: Worm lizard |

Listromycter

This wormlike lizard had a characteristically robust, wedge-shaped skull with simple, peglike teeth and with one tooth centrally positioned in the front of the upper jaw. Living species of this family of burrowers have weak eyesight and legless bodies.

HABITAT Living species of *Listromycter* have a preference for burrowing into loose soil, while actively hunting worms and insects.

Side view of skull

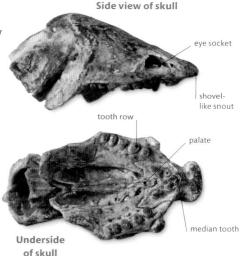

eye socket

shovel-like snout

tooth row

palate

median tooth

Underside of skull

Listromycter leakeyi Charig & Gans; Rusinga Beds; Early Miocene; Kenya.

Typical length 12 in (30 m)

| Range: Early Miocene | Distribution: E. Africa | Occurrence: |

Group: SQUAMATA	Subgroup: VARANIDAE	Informal name: Giant goanna

Varanus

This monitor lizard attained a length twice that of the Recent Komodo Dragon. It had a mandible hinged in the middle, and the widely spaced, recurved teeth were attached to the inner sides of the jaws, with replacements arising between worn teeth. The long neck was made up of nine out of the 29 presacral vertebrae, which have strongly socketed front and bulbous rear articulations.

HABITAT *Varanus* was a ferocious carnivore, living in Australia during the Pleistocene Epoch.
REMARK The hunting technique of this lizard probably included ambushing prey around waterholes and then disabling it by severing its hamstrings.

neural spine

Varanus priscus
Owen; River deposits; Pleistocene; Australia.

rear articulation facet

Vertebra

Typical length 20 ft (6 m)

rear articulation boss

front articulation socket

Range: Pleistocene	Distribution: Australia	Occurrence:

Group: SQUAMATA	Subgroup: MOSASAURIDAE	Informal name: Mosasaur

Prognathodon

Prognathodon was a medium-sized example of the mosasaur family, the largest members of which grew to more than 33 ft (10 m) in length. Like other mosasaurs, its teeth have an almost smooth, recurved crown slightly triangular in cross-section and set on a bulbous root. Individual teeth are located in separate sockets but not attached to the side of the jaw as in other lizards.

HABITAT Mosasaurs were fast-moving marine carnivores, feeding on fish and other marine animals.

Prognathodon sp.;
Upper Chalk; Late Cretaceous; The Netherlands.

recurved tooth crown

smooth enamel

bulbous root of tooth

Jaw fragment with teeth

replacement tooth

Typical length 16½ ft (5 m)

Range: Late Cretaceous	Distribution: Europe	Occurrence:

Group: SQUAMATA	Subgroup: MOSASAURIDAE	Informal name: Mosasaur

Tylosaurus

The mosasaurs had large, lightly built skulls armed with recurved teeth set in sockets and a lower jaw hinged in the middle. Their necks were short, with seven vertebrae, while the remainder of their bodies was elongate. Individual vertebrae had ball-and-socket articulations, allowing considerable sideways movement. The limbs were reduced and paddlelike, used for steering, not propulsion.

HABITAT The mosasaurs evolved from land forms that later returned to an aquatic environment. The group appears to be related to the varanid lizards living today. They actively hunted an assortment of marine organisms in shallow seas.

REMARK Although a short-lived group, appearing and becoming extinct during the Late Cretaceous, the mosasaurs were the most important and widespread of the marine carnivores living at that time. Their remains have been recovered from around the world.

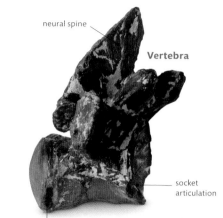

Vertebra

neural spine

socket articulation

bulbous articulation

Tylosaurus nepaeolicus (Cope); Kansas Chalk; Late Cretaceous; US.

Quadrate bone

ulna

radius

finger bones

wrist bones

Forelimb bones

Typical length 20 ft (6 m)

Range: Late Cretaceous	Distribution: N. America	Occurrence:

| Group: SQUAMATA | Subgroup: DINILYSIIDAE | Informal name: Land snake |

Dinilysia

Common features of land snakes are a low, flat braincase, with high mobility of the palate, jaw elements, and quadrate, together with a long temporal region, providing a large attachment area for jaw muscles.

Dinilysia patagonica
Woodward; Rio Colorado
Formation; Cretaceous; Argentina.

HABITAT *Dinilysia* was clearly terrestrial and lived on small vertebrates.

REMARK *Dinilysia*'s skull does not support the suggestion that snakes had a burrowing stage in their ancestry.

long temporal region

flattened skull

eye socket

upper jaw

loose upper/lower jaw articulation

tooth sockets

Skull and lower jaws

Typical length 10 ft (3 m)

| Range: Late Cretaceous | Distribution: S. America | Occurrence: |

| Group: SQUAMATA | Subgroup: PALAEOPHIIDAE | Informal name: Sea snake |

Palaeophis

The vertebrae of this marine snake have socketed front and bulbous rear articulations, with further accessory articulations–characteristic of snakes– situated on the neural arch. The ribs are elongate.

Right lateral view

Left lateral view

bulbous condyle

centrum

HABITAT Like modern sea snakes, *Palaeophis* lived in shallow coastal waters and estuaries.

REMARK Although the vertebrae of *Palaeophis* are relatively common fossil finds, they provide little information about its relationships.

neural spine

Anterior view

cotyle (socket)

Posterior view

neural canal

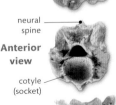

Palaeophis maghrebianus
Arambourg; Early Eocene Phosphates; Morocco.

Dorsal view

Ventral view

Typical length 13 ft (4 m)

| Range: Late Cretaceous–Oligocene | Distribution: Europe, Africa, America | Occurrence: |

Group: SAUROPTERYGIA	Subgroup: LARIOSAURIDAE	Informal name: Pachypleurosaurid

Neusticosaurus

In Permo-Triassic times, certain reptilian groups returned to live in water, which led to a period of reptilian dominance of aquatic habitats. One such group produced the pachypleurosaurs, of which the genus *Neusticosaurus* is an example. The aquatic preference of this genus is indicated by its small, lightly built skull; long, slender body; short limbs; and the simplification and streamlining of its upper limb bones. However, the hands and feet have not evolved into paddles.

HABITAT This marine genus fed on small aquatic organisms.

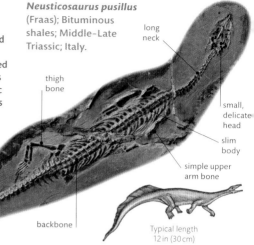

Neusticosaurus pusillus (Fraas); Bituminous shales; Middle–Late Triassic; Italy.

long neck

thigh bone

small, delicate head

slim body

simple upper arm bone

backbone

Typical length 12 in (30 cm)

Range: Middle–Late Triassic	Distribution: Europe	Occurrence:

Group: PLACODONTIA	Subgroup: CYAMODONTIDAE	Informal name: Placodont

Cyamodus

Placodonts were characterized by small heads on short bodies, with stout limbs partly modified into paddles. The skull of this genus is robust and triangular and contains teeth modified into rounded crushing plates. The vertebral centra were strongly biconcave, with long transverse processes. Some forms of this genus developed body armor resembling that of a turtle, but, although aquatic in habit, the modification of the limbs to paddles was only partial.

HABITAT The characteristically large, buttonlike teeth suggest an adaptation for crushing the hard mollusk shells collected in shallow coastal waters.

Dorsal view of skull

Cyamodus munsteri (Agassiz); Muschelkalk; Middle Triassic; Germany.

eye socket

nostril

tooth socket

internal nostril opening

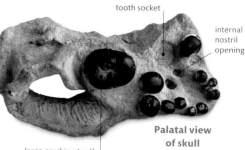

lower jaw articulation

large crushing tooth

Palatal view of skull

Typical length 6½ ft (2 m)

Range: Middle Triassic	Distribution: Europe	Occurrence:

| Group: ICHTHYOPTERYGIA | Subgroup: MIXOSAURIDAE | Informal name: Ichthyosaur |

Mixosaurus

The body of *Mixosaurus* is streamlined and dolphinlike, with a tail that kinks downward. The long-snouted skull is filled with teeth set in sockets, while the limbs are paddle-shaped, with disklike finger elements.

HABITAT Ichthyosaurs were adapted to a marine existence and even gave birth to live young at sea. They were fast-swimming, fish-eating carnivores.

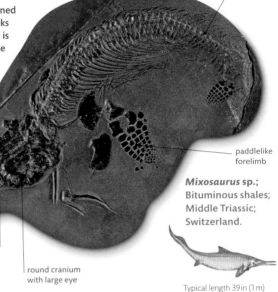

tail

paddlelike forelimb

Mixosaurus sp.;
Bituminous shales;
Middle Triassic;
Switzerland.

nostril

long, multi-toothed snout

round cranium with large eye

Typical length 39 in (1 m)

| Range: Middle Triassic | Distribution: Europe, E. Indies, US, Asia | Occurrence: |

| Group: ICHTHYOPTERYGIA | Subgroup: ICHTHYOSAURIDAE | Informal name: Ichthyosaur |

Ichthyosaurus

Typical ichthyosaur features are: numerous teeth set in a groove within the long jaws; each eye held in a ring of bony plates; the reduction of the pelvic girdle components and associated paddle; a tail kinked downward to support a lower tail lobe; and subdivided finger rows.

HABITAT The ring of strengthening bones around an ichthyosaur's eye indicates it was capable of diving to considerable depths in pursuit of fish.

Ichthyosaurus communis Conybeare; Lower Lias; Early Jurassic; UK.

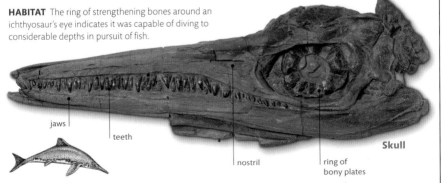

jaws

teeth

nostril

ring of bony plates

Skull

Typical length 6½ ft (2 m)

| Range: Early Jurassic | Distribution: Europe, Greenland | Occurrence: |

Group: ICHTHYOPTERYGIA	Subgroup: OPTHALMOSAURIDAE	Informal name: Ichthyosaur

Platypterygius

Platypterygius had a heavy-beaked, multitoothed head and paddlelike forelimbs with eight to nine digital rows. As in all ichthyosaurs, the vertebral centra are short and biconcave, with no fusion of the neural arch, which has no transverse processes. The double-headed rib articulates via two indistinct bosses on the centrum.

HABITAT *Platypterygius* was a marine reptile, living on fish.

Typical length 16½ ft (5 m)

Platypterygius australis (McCoy); Toolebuc Limestone; Early Cretaceous; Australia.

neural arch attachment

concave vertebral centrum

Four vertebrae

bosses for rib

Range: Early–Late Cretaceous	Distribution: Worldwide	Occurrence:

Group: PLESIOSAURIA	Subgroup: CRYPTOCLIDIDAE	Informal name: Plesiosaur

Cryptoclidus

A medium-sized plesiosaur, *Cryptoclidus* had a small, lightly built skull set on a long neck. The teeth were boomerang-shaped, faintly ribbed, pointed, and round in cross-section. The vertebrae were flat-faced, with two distinct openings (foramina) situated on the undersurface of the centrum. The ribs were single-headed, the girdle elements large and platelike, and the limbs paddle-shaped, with the upper arm bone flared away from the attachment point.

HABITAT The plesiosaurs hunted fish in warm seas.

Cryptoclidus eurymerus (Philips); Oxford Clay; Late Jurassic; UK.

upper arm bone

articulation facets

rounded finger bones

Right forepaddle

modified lower arm bones

Typical length 13 ft (4 m)

Range: Late Jurassic	Distribution: Europe	Occurrence:

Group: PLESIOSAURIA	Subgroup: PLIOSAURIDAE	Informal name: Pliosaur

Liopleurodon

Liopleurodon had a large head; a short neck and tail; and long, paddle-shaped limbs. The vertebral centra of the neck are compressed front to back and are flat-faced, with facets that act as rib articulations. The teeth, which rise from sockets, are long and bow-shaped, with some longitudinal ribbing. Large animals had heads measuring some 10 ft (3 m) in length, and held 12-in (30-cm) long teeth.

HABITAT Pliosaurs ate fish and squidlike organisms (belemnites), which abounded during the Late Jurassic.

Liopleurodon ferox Sauvage; Oxford Clay; Late Jurassic; UK.

Vertebral centrum

flat-faced centrum

rib articulation

Neck rib

base with two foramina

Typical length 33 ft (10 m)

Range: Late Jurassic	Distribution: Europe	Occurrence:

Group: RHYNCHOSAURIA	Subgroup: RHYNCHOSAURIDAE	Informal name: Rhynchosaurid

Hyperodapedon

This rhynchosaur had a broad skull, in which the front of the upper and lower jaws had developed into curved, toothlike, bony processes. The cheek teeth are set in sockets in multiple longitudinal rows.

HABITAT Rhynchosaurs were four-footed, pig-sized vegetarians living in arid regions.
HABITAT These lizardlike reptiles were thought to be related to the living tuatara of New Zealand, but the resemblance is only superficial and they are now placed in a separate order. Their remains are commonly found in Late Triassic rocks.

Lateral view of skull

nostril   eye socket

bony projections

lower jaw

upper jaw

rows of teeth

lower jaw

Basal view of skull

Hyperodapedon gordoni Huxley; Elgin Sandstone; Late Triassic; UK.

Typical length 6½ ft (2 m)

Range: Late Triassic	Distribution: Europe, Asia	Occurrence:

| Group: TESTUDINATA | Subgroup: PROGANOCHELYDAE | Informal name: Primitive turtle |

Proganochelys

Turtles and tortoises are recognizable by their large eye sockets, toothless jaws with horny coverings forming a beak, and bony shells. *Proganochelys* displays its primitiveness in retaining toothlike denticles on the palate, as well as nasal bones. It has an extra series of peripheral plates in the carapace.

large eye socket

complete skull roof

toothless jaws

HABITAT This semiaquatic form fed on plants.

Typical length 39 in (1 m)

Proganochelys quenstedti Baur; Upper Stubensandstein (Keuper); **Late Triassic; Germany.**

| Range: Late Triassic | Distribution: Europe, S. Asia | Occurrence: |

| Group: TESTUDINATA | Subgroup: PELOMEDUSIDAE | Informal name: Side-necked turtle |

Pelusios

The ability to withdraw the head sideways into the shell, the possession of a mesoplastral element not restricted to the lower shell's outer margin, and the firm attachment of the rear girdle to the lower shell are all features of the pelomedusid family. The first side-necked turtle appeared in the Early Cretaceous of Africa, but by the Tertiary, they had a nearly worldwide distribution.

anterior

mesoplastral element

entoplastron

hinge

HABITAT All pelomedusids appear to have been semiaquatic omnivores. Living species are now restricted to the southern continents of Africa and South America.

Internal view

rear girdle scars

Pelusios sinuatus (Smith); Olduvai Beds; Early Pleistocene; Tanzania.

External view

Typical length 20 in (50 cm)

| Range: Early Pleistocene–Recent | Distribution: Africa, S. America | Occurrence: |

Group: CRYPTODIRA	Subgroup: MEIOLANIIDAE	Informal name: Horned tortoise

Meiolania

The bizarre head of this tortoise could not be retracted into its shell because of its large size and the bony spikes that adorn the skull roof. The tail was heavily armored and clublike, an adaptation that probably helped when defending its territory.

HABITAT *Meiolania* belonged to a family of tortoises that were omnivorous and lived on land.

REMARK It is believed that this family may have had its origins in the Late Cretaceous of South America, as a similar genus occurred there at an earlier period.

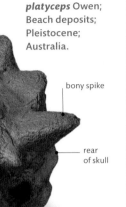

Skull

Meiolania platyceps Owen; Beach deposits; Pleistocene; Australia.

bony spike

rear of skull

auditory opening of cranium

eye socket

Typical length 6½ ft (2 m)

Range: Pleistocene	Distribution: Australia	Occurrence:

Group: TESTUDINATA	Subgroup: TRIONYCHIDAE	Informal name: Mud turtle

Trionyx

The bony shell of *Trionyx* can be distinguished from other turtles' shells by its lack of peripheral plates, highly textured surface, lack of horny outer plates, and loosely connected plastron (ventral shell).

HABITAT All trionychids were aquatic omnivores and are still found living in lakes, estuaries, and slow-moving rivers. In the Early Tertiary, the larger family members approached 6½ ft (2 m) in length.

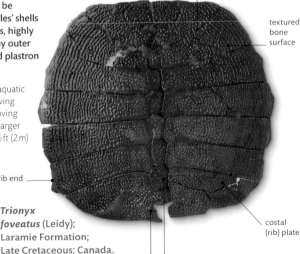

interconnecting suture

textured bone surface

rib end

costal (rib) plate

Trionyx foveatus (Leidy); Laramie Formation; Late Cretaceous; Canada.

front | neural plate

Typical length 36 in (90 m)

Range: Cretaceous–Recent	Distribution: Asia, Africa, N. America, Europe	Occurrence:

| Group: TESTUDINATA | Subgroup: CHELONIIDAE | Informal name: Marine turtle |

Puppigerus

This genus has characteristic adaptations for an aquatic existence: large eye sockets, with a secondary skull roof behind them, and a fully formed secondary palate. The latter was especially important, as it prevented an unwanted intake of water while feeding below the surface. In adults, the shell is fully ossified, with an outer margin of peripheral plates, while the plastron (ventral shell) comprises four paired elements and one centrally placed, all of which are loosely connected by fingerlike projections in the midline. The forelimbs developed into flippers.

HABITAT *Puppigerus* was a marine turtle and, like its modern counterparts, fed on sea grasses. Its hindlimbs were less flipperlike, a condition that possibly indicates a greater mobility on land.
REMARK Like modern sea turtles, females would have buried their eggs on sandy beaches.

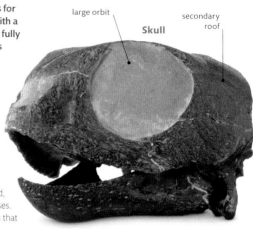

large orbit

Skull

secondary roof

Puppigerus crassicostata
(Owen); London Clay; Early Eocene; UK.

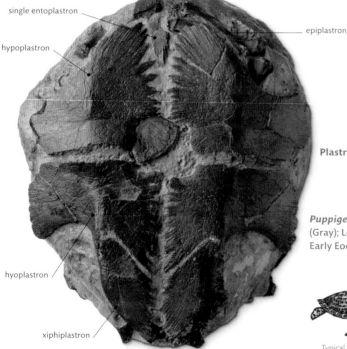

single entoplastron

epiplastron

hypoplastron

Plastron

Puppigerus camperi
(Gray); London Clay; Early Eocene; UK.

hyoplastron

xiphiplastron

Typical length 36 in (90 m)

| Range: Eocene | Distribution: Europe | Occurrence: |

Group: TESTUDINATA	Subgroup: PROSTEGIDAE	Informal name: Marine turtle

Cimochelys

Cimochelys benstedi
(Owen); Middle Chalk;
Late Cretaceous; UK.

The low-profiled carapace of this small turtle is not completely ossified. This feature is clearly seen between the ends of each expanded rib and the marginal plates and is characteristic of some species of marine turtles. The central row of plates (neurals) are distinctly keeled in this genus.

HABITAT *Cimochelys* was a marine vegetarian.

REMARK Being very agile in water, marine turtles are less vulnerable to attack. So in many forms, the bony armor of the shell has become reduced.

expanded rib

reduced carapace

keeled neural

Typical length 12 in (30 cm)

Range: Late Cretaceous	Distribution: Europe	Occurrence:

Group: TESTUDINATA	Subgroup: TESTUDINIDAE	Informal name: Tortoise

Geochelone

Geochelone pardalis
(Bell); Laetolil Beds;
Pliocene; Tanzania.

A common characteristic of tortoises like *Geochelone* is a high-arched, fully ossified, bumpy shell embellished with concentric markings. Other characteristic features of this land-dwelling group include a skull that lacks a secondary bone covering behind the eye sockets and the ability to withdraw the head vertically and backward.

HABITAT Like many tortoises, this genus lives in arid areas, feeding on a variety of plant matter.

high-arched shell

lower shell

marginal plates

Typical length 14 in (35 cm)

Range: Eocene–Recent	Distribution: Europe, Asia, Africa	Occurrence:

| Group: ORNITHOSUCHIA | Subgroup: ORNITHOSUCHIDAE | Informal name: Ornithosuchid |

Riojasuchus

The head of *Riojasuchus* is large but lightly built, with the upper jaw curved down over the lower. There is a skull opening just before the eye socket, and the teeth are large, bladelike, compressed sideways, and recurved. The hip joint is only partly open, and the head of the thigh bone only slightly turned in. There are three pairs of sacral ribs.

HABITAT All ornithosuchids were land dwellers, mainly four-footed, and living on a flesh diet.

REMARK This family is a member of the Archosauria, which includes dinosaurs, pterosaurs, and crocodiles.

Riojasuchus tenuisceps
Bonaparte; Los Colorados
Formation; Late Triassic;
Argentina.

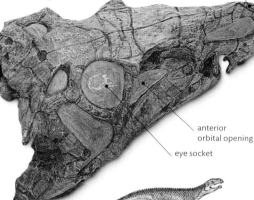

downcurved upper jaw

anterior orbital opening

eye socket

Typical length
10 ft (3 m)

| Range: Late Triassic | Distribution: S. America | Occurrence: |

| Group: PHYTOSAURIA | Subgroup: PHYTOSAURIDAE | Informal name: Phytosaurid |

Belodon

Outwardly, the body of this phytosaur resembled that of our modern, long-snouted crocodiles, having short legs, an ornate sculptured surface to the skull, and considerable body armor. In detail, however, the ankle joint and the platelike elements of the hindlimb girdle indicate a more primitive condition, and thus a more awkward gait. The nostrils being situated on top of the head just forward of the eye sockets, rather than at the tip of the snout as in crocodiles, provides an instant identification feature. These nostril openings are also elevated above the other skull bones so that they could protrude above the water when the remainder of the body was submerged.

Belodon plieningeri
Meyer; Keuper
Sandstone; Late
Triassic; Germany.

HABITAT The long snout suggests a fish eater, but preserved stomach contents show *Belodon* fed upon a variety of reptiles.

nostril openings

eye socket

long, hooked snout

tooth row

jaw articulation

Typical length 10 ft (3 m)

| Range: Late Triassic | Distribution: Europe | Occurrence: |

Group: CROCODYLIFORMES	Subgroup: METRIORHYNCHIDAE	Informal name: Marine crocodile

Gracilineustes

The metriorhynchid family is the most specialized of all known crocodiles and is perhaps the only archosaur group to become fully aquatic. Changes in its original habit are clearly reflected in its skeleton, as the forelimbs were transformed into paddles, the neck was shortened, the tail bent downward at the end to support a large caudal fin, and the body armor lost. In contrast to their more terrestrial crocodilian cousins, the skull is long and lightly built—this was another requirement for a fully aquatic mode of life.

HABITAT Species of metriorhynchids were particularly common in the Jurassic seas of Europe, where they hunted for fish and squidlike animals that shared the same habitat. It may be that they only came to land to lay their eggs, in the manner of modern turtles. It is also possible that they hauled themselves onto sand banks to bask after hunting for fish.
REMARK The metriorhynchids belong to the primitive mesosuchian suborder of crocodiles, which appeared in the Triassic but finally became extinct in the earliest part of the Tertiary, some 60 million years ago.

nostril

long snout

Gracilineustes leedsi (Andrews); Oxford Clay; Late Jurassic; UK.

neural spine

articulation

flat-faced articulation

Vertebra

eye sockets

plain bone surface

large temporal opening

jaw articulation

Skull

Typical length 10 ft (3 m)

Range: Middle–Late Jurassic	Distribution: Europe, S. America	Occurrence:

Group: CROCODYLIFORMES	Subgroup: GONIOPHOLIDIDAE	Informal name: Crocodile

Goniopholis

Goniopholids had strongly built, low-profiled skulls, with ornamented bone surfaces and patterned body armor, thus differing from the metriorhynchids. Like their marine cousins, however, they had nearly flat-faced centra to their vertebrae, which contrasts with the ball-and-socket development found in the more modern forms.

HABITAT Like most unspecialized crocodiles, *Goniopholis* lived a semiaquatic existence, feeding upon a mixture of animal and plant matter.

Goniopholis crassidens Owen; Wealden Beds; Early Cretaceous; UK.

Dorsal armor

rib
fragment | matrix | highly ornate
bony scute

Typical length
10 ft (3 m)

Range: Late Jurassic–Late Cretaceous	Distribution: Europe, Asia, N. America	Occurrence:

Group: CROCODYLIFORMES	Subgroup: GAVIALIDAE	Informal name: Gavial

Rhamphosuchus

Modern crocodiles are divided into three families: alligators, crocodiles, and gavials. *Rhamphosuchus* is a member of the gavial family. The Gavialidae are regarded as specialized fish eaters, catching their prey with the aid of long, sharply pointed teeth situated in long, slender jaws. In *Rhamphosuchus*, however, the teeth are robust and conical, which is suggestive of a mixed diet.

Rhamphosuchus crassidens Falconer & Cautley; Siwalik Series; Pliocene; Pakistan.

HABITAT This giant was an inhabitant of the larger rivers and lakes of Central Asia.

elongate snout   nostrils

conical teeth

Typical length
50 ft (15 m)

Range: Pliocene	Distribution: Asia	Occurrence:

Group: CROCODYLIFORMES	Subgroup: DIPLOCYONODONTIDAE	Informal name: Alligator

Diplocynodon

Alligators like the medium-sized *Diplocynodon* can be distinguished from true crocodiles by the absence of a pit to house the fourth tooth of the lower jaw. In other respects, the skull anatomy of the modern families is basically similar, having a strongly buttressed skull with a highly ornate bony surface, a jaw articulation set well back to facilitate the wide opening of the mouth, and a variety of rounded to sharply pointed teeth placed in deep sockets in the jaws. The remainder of the crocodilian skeleton is even more conservative, as its basic plan has been unchanged since the Triassic.

HABITAT Unlike Recent species, *Diplocynodon* had a distribution that encompassed both North America and Europe. Although alligators are now restricted to America, this genus was particularly common in the swamps of Europe in Oligocene times. It had a mixed diet of animal and plant matter.

REMARK The crocodiles survived the great extinction at the end of the Mesozoic, becoming even more numerous and widespread during the warmer periods of the Tertiary. But their more primitive cousins became extinct at the start of this era.

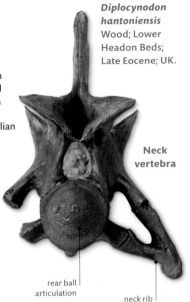

Diplocynodon hantoniensis
Wood; Lower Headon Beds; Late Eocene; UK.

Neck vertebra

rear ball articulation

neck rib

eye socket

broad snout

Skull

nostril

Typical length 10 ft (3 m)

Range: Eocene–Pliocene	Distribution: Europe, N. America	Occurrence:

Group: PTEROSAURIA	Subgroup: PTERODACTYLIDAE	Informal name: Pterosaur

Diopecephalus

Diopecephalus is thought to have been a small-toothed insectivore, belonging to an order of archosaurs characterized by: delicate skulls with lightly built skeletons; paper-thin, hollow bones; greatly extended first fingers, supporting a wing membrane; and short legs. All family members had short tails and long skulls.

Diopecephalus kochi (Wagner); Solnhofen Limestone; Late Jurassic; Germany.

HABITAT The pterosaurs were the first vertebrates to develop powered flight, chasing and catching their prey on the wing, in a manner similar to that of certain birds.

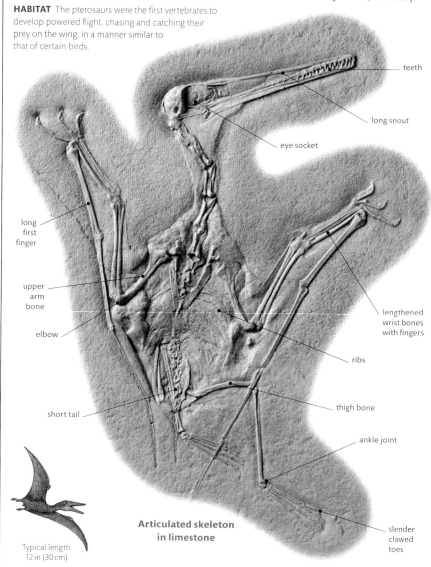

teeth

long snout

eye socket

long first finger

upper arm bone

elbow

lengthened wrist bones with fingers

ribs

thigh bone

short tail

ankle joint

Typical length 12 in (30 cm)

Articulated skeleton in limestone

slender clawed toes

Range: Late Jurassic	Distribution: Europe, Asia, US, Africa	Occurrence:

DINOSAURS

THIS SUCCESSFUL GROUP, made up mainly of land-dwelling reptiles, first appeared about 225 million years ago, having evolved from early archosaurs. They are divided into two groups based on their skeletal anatomy. These were the herbivorous Ornithischia or "bird-hipped" dinosaurs and the Saurischia or "lizard-hipped" dinosaurs. This latter group comprises the large four-footed Sauropods (Sauropodomorphs) plus the two-footed, flesh-eating Theropods that gave rise to the birds. The remarkable success of the dinosaurs has been attributed to the modifications in the limb and girdle bones, which so improved their stance and gait that they were able to adapt to new habitats. Their sudden extinction in the Late Cretaceous ended 150 million years of domination of the land. However, the birds, derived from small therapods, can be referred to as "avian dinosaurs" and survived and lived on to dominate the skies.

Group: THEROPODA	Subgroup: COMPSOGNATHIDAE	Informal name: Compsognathid

Compsognathus

This small, chicken-sized dinosaur had recurved, serrated teeth set in a lightly built skull. The limbs are long, slender, and hollow and the pubic bone faces forward.

HABITAT This highly mobile, bipedal carnivore preyed upon insects and small, lizardlike creatures, which it stalked and snatched from the ground.
REMARK Its hunting habit, therefore, could have been in direct competition with those suggested for the similar-sized bird *Archaeopteryx*.

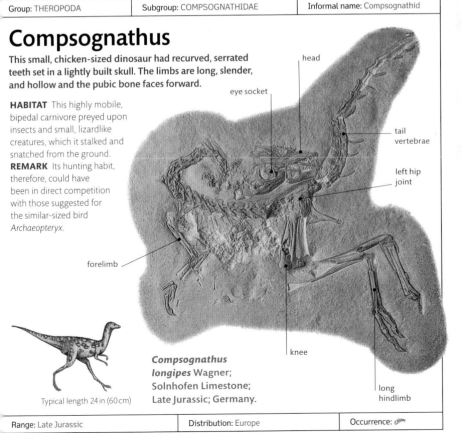

head

eye socket

tail vertebrae

left hip joint

knee

forelimb

long hindlimb

Compsognathus longipes Wagner; Solnhofen Limestone; Late Jurassic; Germany.

Typical length 24 in (60 cm)

Range: Late Jurassic	Distribution: Europe	Occurrence:

Group: THEROPODA	Subgroup: PROCERATOSAURIDAE	Informal name: Ceratosaurid

Proceratosaurus

This medium-sized dinosaur is known only from a single, lightly built skull. The skull has a nasal horn and jaws filled with sharply pointed, recurved, and serrated teeth.

Proceratosaurus bradleyi (Woodward); Greater Oolite; Middle Jurassic; UK.

HABITAT Despite the limited remains, it is possible to say with some certainty that *Proceratosaurus* was an agile, bipedal predator, which was capable of overpowering slower-moving reptiles.

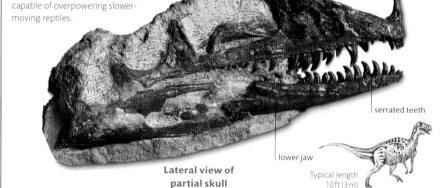

nasal horn

serrated teeth

lower jaw

Lateral view of partial skull

Typical length 10 ft (3 m)

Range: Middle Jurassic	Distribution: Europe	Occurrence:

Group: THEROPODA	Subgroup: Tyrannosauridae	Informal name: Tyrannosaurid

Daspletosaurus

A large head and powerful jaws, together with shortened forelimbs bearing only two fingers, are typical features of this family of carnivorous dinosaurs.

HABITAT *Daspletosaurus* and its close relatives were probably ferocious killers of herbivorous dinosaurs, as well as being scavengers of carrion.

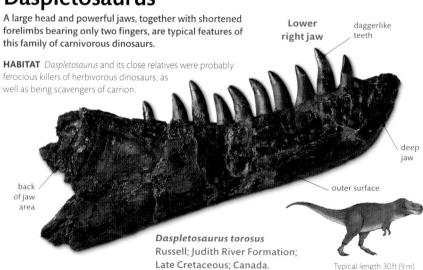

Lower right jaw

daggerlike teeth

deep jaw

outer surface

back of jaw area

Daspletosaurus torosus Russell; Judith River Formation; Late Cretaceous; Canada.

Typical length 30 ft (9 m)

Range: Late Cretaceous	Distribution: N. America	Occurrence:

| Group: THEROPODA | Subgroup: ARCHAEOPTERYGIDAE | Informal name: Urvogel |

Archaeopteryx

This small, chicken-sized therapod dinosaur was, for more than a century, regarded as a link between birds and dinosaurs or the first true bird. It possesses a number of reptilian features. For example, the lightly built skull has true teeth set in sockets in the jaws; the breast bone is small and lacks a keel; the forelimb skeleton retains three functional fingers; and there is a long, bony tail. *Archaeopteryx* also possesses characters that, until recently, one would only associate with birds, including reduced fingers; a wishbone (furcula); and, above all, feathers. The discovery of feathers and featherlike structures on theropod dinosaurs, as well as the discovery of flying dinosaurs in China, has cast doubt on whether *Archaeopteryx* is a direct ancestor of the birds (avian dinosaurs) or just closely related to their origin.

HABITAT *Archaeopteryx* hunted fish and insects along the arid shorelines of subtropical coastal lagoons. It is likely that it could climb trees and to a limited extent was able to fly in order to evade predators.

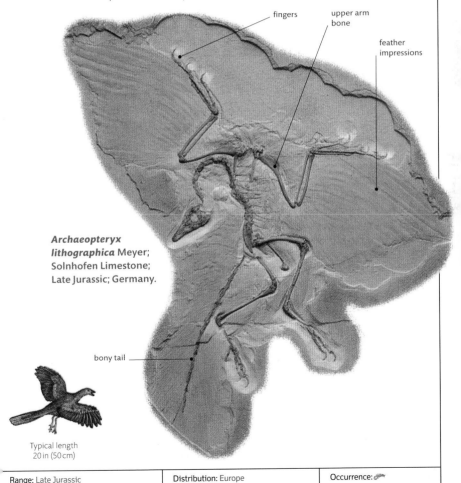

fingers

upper arm bone

feather impressions

Archaeopteryx lithographica Meyer; Solnhofen Limestone; Late Jurassic; Germany.

bony tail

Typical length
20 in (50 cm)

| Range: Late Jurassic | Distribution: Europe | Occurrence: |

Group: THEROPODA	Subgroup: ORNITHOMIMIDAE	Informal name: Ostrich dinosaur

Gallimimus

The ostrich dinosaurs were long-limbed, lightly built, small-headed reptiles, with a tendency to become toothless. In *Gallimimus*, the toothless jaws were covered by a horny beak.

HABITAT Although presumed to be a carnivore, it has been suggested that *Gallimimus* also raided the nests of other reptiles for their eggs.

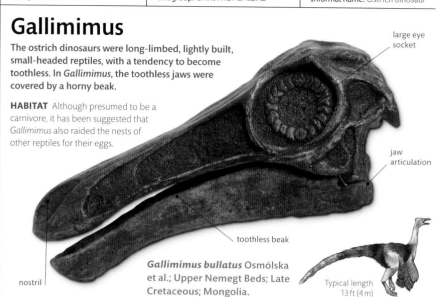

large eye socket

jaw articulation

toothless beak

nostril

Gallimimus bullatus Osmólska et al.; Upper Nemegt Beds; Late Cretaceous; Mongolia.

Typical length 13 ft (4 m)

Range: Late Cretaceous	Distribution: Asia	Occurrence:

Group: SAUROPODOMORPHA	Subgroup: CETIOSAURIDAE	Informal name: Sauropod

Cetiosaurus

The small head, long neck and tail, and solid limb bones of this giant show it to be a member of the reptile-footed dinosaurs. Its banjolike vertebrae are characteristic.

HABITAT *Cetiosaurus* was a plant eater, but it is not certain whether it inhabited the margins of lakes and rivers or was a plains wanderer. The latter suggestion is the most favored at present.

Vertebra

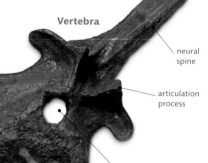

neural spine

articulation process

neural canal

flat-faced centrum

Cetiosaurus leedsi (Hulke); Oxford Clay; Middle Jurassic; UK.

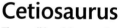

Typical length 65 ft (20 m)

transverse process

Range: Middle–Late Jurassic	Distribution: Europe	Occurrence:

Group: SAUROPODOMORPHA	Subgroup: DIPLODOCIDAE	Informal name: Sauropod

Diplodocus

This large, herbivorous quadruped had a long neck and tail and a small, elongate head. The rakelike teeth, situated toward the front of the jaws, together with the short front legs and the beamlike chevron bones under the tail vertebrae, are characteristic.

HABITAT *Diplodocus* probably lived close to areas of fresh water, using its specially adapted teeth to gather in surrounding vegetation.

REMARK An adult probably weighed up to twice as much as a bull African elephant.

Diplodocus longus Marsh; Morrison Formation; Late Jurassic; US.

Tail vertebra

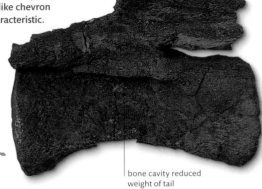

bone cavity reduced weight of tail

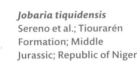

Typical length 88½ft (27 m)

Range: Late Jurassic	Distribution: N. America	Occurrence:

Group: SAUROPODOMORPHA	Subgroup: MACRONARIA	Informal name: Sauropod

Jobaria

The basic body shape of *Jobaria* resembled that of other sauropods, but the neck and tail were somewhat shorter. The skull was characteristically flat-faced with a round profile, and the leaf-shaped teeth extended along the length of the jaws.

HABITAT Like others of its type, this genus was a plant-eating quadruped which lived close to well-watered areas. The large teeth were well adapted for shearing vegetation.

Jobaria tiquidensis Sereno et al.; Tiouarén Formation; Middle Jurassic; Republic of Niger.

cutting edge

Leaf-shaped tooth

root

tooth

Fragment of lower jaw with teeth

Typical length 65 ft (20 m)

Range: Early Cretaceous	Distribution: Africa	Occurrence:

Group: SAUROPODOMORPHA	Subgroup: TITANOSAURIDAE	Informal name: Titanosaur

Neuquensaurus

Tail vertebra

The titanosaurs were the last of the reptile-footed plant eaters, but their remains are usually fragmentary. However, they appear to have followed the type in having small heads, long tails, and elephantine limbs, although the neck was shorter. Their most important distinguishing feature is the ball-and-socket articulation of the vertebrae at the beginning of the tail. Being long-faced, the most complete skull is said to resemble that of *Diplodocus*. At least one species developed bony scutes as armor.

HABITAT Like others of its type, it may not have wandered far from water.

ball articulation

Typical length 40 ft (12 m)

Neuquensaurus australis (Lydekker); Late Cretaceous; Argentina.

socket articulation

Range: Cretaceous	Distribution: Europe, Africa, Asia, S. America	Occurrence:

Group: ORNITHISCHIA	Subgroup: HETERODONTOSAURIDAE	Informal name: Heterodontosaurid

Heterodontosaurus

Heterodontosaurus tuckii Crompton & Charig; Cave sandstone; Early Jurassic; South Africa.

The lower jaws of this early bipedal genus were toothless in front—a feature common to the order. However, it is unusual in that two pairs of fanglike teeth accompany the single-rowed, shearing dentition of the cheek region. The presence of teeth at the front of the upper jaw is regarded as primitive.

HABITAT This land-living plant eater may also have grubbed out roots.

eye-socket bar

Skull

shearing teeth

fanglike canine

Typical length 39 in (1 m)

Range: Early Jurassic	Distribution: S. Africa	Occurrence:

| Group: ORNITHISCHIA | Subgroup: HYPSILOPHODONTIDAE | Informal name: Hypsilophodontid |

Hypsilophodon

The hypsilophodonts were medium-sized, light-bodied, land-dwelling bipeds. They show their primitiveness in having teeth situated in the front of the upper jaws and by the extreme length of the backward-facing pubic rod. The cheek teeth were of the shearing type, arranged in a single row and replaced in batches of three. The feet were four-toed, each terminating in a pointed hoof.

HABITAT *Hypsilophodon* fed upon the fern and cycadlike plants that dominated the early Cretaceous.

socket articulation

Single toe

tendon attachment

pointed hoof

ball articulation

Hypsilophodon foxii
Huxley; Wessex Formation;
Early Cretaceous; UK.

Typical length 8 ft (2.5 m)

| Range: Late Jurassic–Early Cretaceous | Distribution: Europe, N. America | Occurrence: |

| Group: ORNITHISCHIA | Subgroup: IGUANODONTIDAE | Informal name: Iguanodontid |

Iguanodon

This genus is made up of large, heavily built, land-dwelling semibipeds. The skull was rather long in profile and lacked teeth at the front of the beaklike jaws. The cheek teeth were arranged in a single row and were leaf-shaped when new and chisel-shaped when worn. The forelimb was heavy, with the first digit of the hand ending in a spike.

HABITAT *Iguanodon* was common in Europe during the early part of the Cretaceous. It fed on plants.

muscle attachment area

single-rowed, leaflike teeth

Lower jaw

inner surface

Iguanodon sp.; Weald Clay Group; Early Cretaceous; UK.

front of jaw

Typical length 30 ft (9 m)

| Range: Early Cretaceous | Distribution: Europe, Asia, Africa, N. America | Occurrence: |

Group: ORNITHISCHIA	Subgroup: HADROSAURIDAE	Informal name: Duck-billed dinosaur

Edmontosaurus

Edmontosaurus annectens
(Marsh); Lance Formation;
Late Cretaceous; US.

The body form of this duck-billed dinosaur was similar to that of the iguanodontids, being large and heavily built. However, the skull was much flatter in profile and the diamond-shaped teeth were arranged in batteries, with as many as 700 being visible at any one time. The tail, too, was more laterally flattened but retained the bony tendon support.

battery of teeth

HABITAT *Edmontosaurus* was a plant eater that lived in large herds in and around lowland forests.

Lower right jaw

front of jaw

Typical length
42½ ft (13 m)

Range: Late Cretaceous	Distribution: N. America	Occurrence:

Group: ORNITHISCHIA	Subgroup: HADROSAURIDAE	Informal name: Duck-billed dinosaur

Parasaurolophus

This large, semibipedal, crested duckbill is identified by the bizarre extension of the nasal bones over the skull, which forms a tubular crest over 5 ft (1½ m) long.

Skull and lower jaws

tubular crest

HABITAT *Parasaurolophus*, like all hadrosaurs, was a grazing herbivore.
REMARK The hadrosaurian crest may have been used for visible identification between species, while the crest's tubular construction could have been used as a resonator to make warning calls to other grazing herbivores.

Parasaurolophus walkeri
Parks; Judith River Formation;
Late Cretaceous; Canada.

eye socket

jaw articulation

nostril

cutting edge of teeth

Typical length
33 ft (10 m)

Range: Late Cretaceous	Distribution: N. America	Occurrence:

Group: ORNITHISCHIA	Subgroup: STEGOSAURIDAE	Informal name: Plated dinosaur

Stegosaurus

This heavy-bodied, quadrupedal genus can be recognized by its very small and low-profiled skull and by the double row of alternating, upward-pointing, angular plates and four tail spikes that adorned the body. The combination of blunt, leaflike teeth and particularly short front legs (about half the size of the rear) also assists in identification.

HABITAT *Stegosaurus* was a plant eater, using its tail spikes only for defense.
REMARK The orientation and alternating pattern of the angular back plates would have made them ideal for regulating the animal's temperature, by radiating heat from its massive body.

Angular, bony plate

articulation facet

Stegosaurus sp.;
Kimmeridge Clay ;
Late Jurassic; UK.

Typical length 30 ft (9 m)

Range: Late Jurassic	Distribution: N. America, Europe	Occurrence:

Group: ORNITHISCHIA	Subgroup: PACHYCEPHALOSAURIDAE	Informal name: Bone-headed dinosaur

Stegoceras

Thickened skull roof

The thickened skull roof, fashioned into a solid, bony dome and partly surrounded by a lumpy frill, distinguishes the bone-headed dinosaurs from the other bipedal ornithopods. Further characteristics include the presence of teeth in the front of the upper jaw, a single row of leaf-shaped cheek teeth, and an excessively shortened pubic rod, which lacks a connecting process to the ischium (pelvic bone).

bony frill

HABITAT All members of the family were herbivores, living in open country.

teeth

eye socket

jaw articulation

Stegoceras validus
Lambe; Judith River
Formation; Late
Cretaceous; Canada.

Typical length 8 ft (2.5 m)

Range: Late Cretaceous	Distribution: N. America	Occurrence:

| Group: ORNITHISCHIA | Subgroup: ANKYLOSAURIDAE | Informal name: Armored dinosaur |

Euoplocephalus

This tanklike quadruped can be identified by its skull, protected by a series of scutes fused to its surface, and by its body, covered by a mosaic of flat and keel-shaped, interlocking, bony plates which narrow to a tail that terminates in a huge club.

HABITAT This genus probably lived in arid areas.

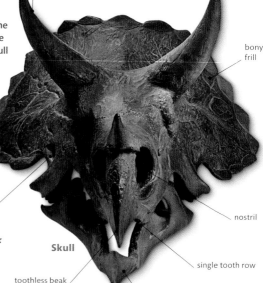

End of tail

vertebrae

mummified skin

bony club

Typical length
20 ft (6 m)

Euoplocephalus tutus
(Lambe); Horseshoe Canyon
Formation; Late Cretaceous;
Canada.

| Range: Late Cretaceous | Distribution: N. America | Occurrence: |

| Group: ORNITHISCHIA | Subgroup: CERATOPSIDAE | Informal name: Horned dinosaur |

Triceratops

A snout fashioned into a laterally compressed beak, with the upper jaw overhanging the lower, is a unique feature of the horned dinosaurs. *Triceratops* exemplifies the more advanced forms, having three bony horns, with the back of the skull extended into a huge bony frill.

HABITAT This was a large, quadrupedal herbivore, which used its powerful beak to slice through the tougher stems of plants. It may have lived in more open habitats, forming small herds in a similar fashion to many present-day plains mammals.

Triceratops prorsus Marsh;
Lance Formation; Late
Cretaceous; US.

lateral bony horn

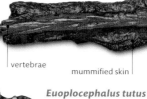

bony frill

eye socket

nostril

Skull

single tooth row

toothless beak

lower jaw

Typical length 30 ft (9 m)

| Range: Late Cretaceous | Distribution: N. America | Occurrence: |

Group: ORNITHISCHIA	Subgroup: PROTOCERATOPSIDAE	Informal name: Horned dinosaur

Protoceratops

This small, stocky quadruped is characterized by a horny beak, a small nose horn, leaflike cheek teeth set in a single line, and a broad neck frill containing large perforations.

HABITAT *Protoceratops* probably lived in open areas, where it fed upon plant stems which it cut off with its powerful beak. The stems were then sheared into smaller pieces by the cheek teeth.

REMARK In the 1920s, an American expedition to Mongolia discovered numerous dinosaur nests with eggs deliberately arranged in neat, concentric rings. Some eggs contained bones of partly developed young, thought to initially belong to *Protoceratops*. However, the eggs were later proved to belong to oviraptors. More recently, a nest with 15 infant *Protoceratops* was discovered in Mongolia, suggesting that parental care was present long after hatching.

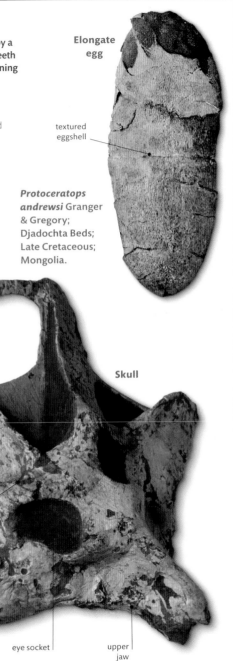

Elongate egg

textured eggshell

Protoceratops andrewsi Granger & Gregory; Djadochta Beds; Late Cretaceous; Mongolia.

frill

Skull

skull roof

eye socket

upper jaw

Typical length 6ft (1.8m)

Range: Late Cretaceous	Distribution: Asia	Occurrence:

BIRDS

IT IS NOW WIDELY ACCEPTED that some nonavian dinosaurs were capable of flight and that the acquisition of birdlike characteristics was gradual and progressive. It is difficult to define the point in time when birds (avian dinosaurs) first arose; the general consensus is that it was during the Mid- to Late Jurassic.

Group: HESPERORNITHIFORMES	Subgroup: HESPERORNITHIDAE	Informal name: Toothed bird

Hesperornis

These birds were specialized, flightless, foot-propelled divers. In appearance, their bodies resembled those of modern divers and grebes, but there is no actual relationship. The skeleton possessed a skull with true teeth; a keel-less breast bone; a much-reduced forelimb; and some characteristically avian, saddle-shaped vertebral centra. The aquatic modifications of *Hesperonis* included thick-walled and nonpneumatic bone and laterally flattened lower leg bones.

HABITAT *Hesperonis* was a fish-eating carnivore, living in warm seas.

Hesperornis regalis Marsh; Kansas Chalk; Late Cretaceous; US.

ankle joint

Fused and flattened metatarsal bones

Typical length 5 ft (1.5 m)

Saddle-shaped centrum

Range: Late Cretaceous	Distribution: N. America	Occurrence:

Group: CARIAMIFORMES	Subgroup: PHORUSRHACIDAE	Informal name: Terror bird

Phorusrhacos

The most outstanding feature of this large, flightless genus is its gigantic hooked bill, which resembles that of a powerful eagle. It has been suggested that it is related to the present-day seriemas and falcons.

Skull

eye socket   nostril

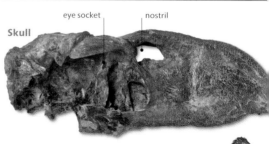

HABITAT It probably hunted over open plains country.
REMARK The family first appeared during the Oligocene in South America, where they evolved and became the dominant carnivore. The superiority lasted until about 4 million years ago, when they became extinct before the end of the Pliocene.

Phorusrhacos inflatus Ameghino; Santa Cruz Formation; Miocene; Argentina.

Typical length 5 ft (1.5 m)

Range: Miocene	Distribution: S. America	Occurrence:

| Group: ODONTOPTERYGIDIFORMES | Subgroup: PELAGORNITHIDAE | Informal name: Bony-toothed bird |

Dasornis

The bony-toothed birds were long-winged seabirds, with unique toothlike projections along the cutting edge of the jaws. In the diminutive *Dasornis*, these projections have a distinct forward slant and are set in a 6 in (15 cm) long skull, similar in form and size to that of the living gannet. The larger species are thought to have had a wingspan of about 16½ ft (5 m).

HABITAT *Dasornis* was certainly a fish eater, perhaps snatching its prey from the sea as it glided over the surface.

REMARK The bony-toothed birds were among the largest flying birds ever to have lived.

Dasornis toliapica (Owen); London Clay; Early Eocene; UK.

Skull

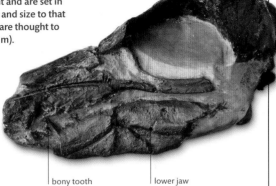

bony tooth

lower jaw

Typical length 36 in (90 cm)

| Range: Early Eocene | Distribution: Europe | Occurrence: |

| Group: AEPYORNITHIFORMES | Subgroup: AEPYORNITHIDAE | Informal name: Elephant bird |

Aepyornis

Aepyornis was the largest genus of this ostrichlike family, popularly known as the elephant birds. It grew to an estimated weight of 1,000 lb (450 kg)—by far the heaviest bird ever to have existed. Its weight prevented any form of flight, resulting in the loss of the keel on the breast bone and the reduction of the wings. The legs, however, were relatively short and powerful, with a three-toed foot. The ovoid eggs of these birds, found buried in sand dunes, were huge, the largest examples having a liquid capacity of 2⅓ gallons (8.5 liters).

HABITAT Elephant birds are thought to have been browsers, feeding on fruit and leaves.

ankle joint

Egg

Aepyornis maximus Geoffroy; Superficial deposits; Quaternary; South Madagascar.

toe articulations

Massive metatarsus

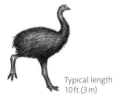

Typical length 10 ft (3 m)

| Range: Pleistocene–Recent | Distribution: Madagascar | Occurrence: |

Group: SPHENISCIFORMES	Subgroup: SPHENISCIDAE	Informal name: Penguin

Pachydyptes

This very large genus of penguin was similar in structure to modern forms, its wings having developed into flippers for swimming. The modifications to the forelimb included a marked flattening of the individual elements, the pronounced development of the head end of the upper arm bone, and a considerable thickening of the bone walls. Except for the excessively short and broad tarsus, the remainder of the skeleton was similar to those of other aquatic, wing-propelled divers.

HABITAT All penguins are marine and feed on fish caught underwater.

Pachydyptes ponderosus
Oliver; Runangan stage; Late Eocene; New Zealand.

Typical length
4½ ft (1.3 m)

enlarged head

flattened shaft

Humerus front and side view

compressed articulation

elbow joint

Range: Late Eocene–Early Oligocene	Distribution: New Zealand	Occurrence:

Group: PROCELLARIIFORMES	Subgroup: PROCELLARIIDAE	Informal name: Cory's shearwater

Calonectris

This is a large genus, distinguished from other similar-sized shearwaters by its relatively short-hooked bill, more rounded wing bones, less forwardly projecting keel to the breast bone, and proportionately longer tarsus.

HABITAT As the Cory's shearwater favors the warmer waters, its breeding range now extends no farther north than the Mediterranean Sea and the Atlantic seaboard of Portugal. However, 100,000 years ago, it is known to have inhabited certain sea caves of south Wales, where remains of nestlings and adults have been found.

Associated skeleton

Calonectris diomedea
(Scopoli); Cave deposits; Pleistocene; Gibraltar.

Typical length
16 in (40 cm)

upper arm bone

skull

Range: Pleistocene–Recent	Distribution: Mediterranean, Atlantic Ocean	Occurrence:

| Group: APODIFORMES | Subgroup: AEGIALORNITHIDAE | Informal name: Swift |

Primapus

The upper arm bone of this genus exhibits structures clearly adapted for a specialized aerial mode of life. For example, it is short and stout, the head is highly developed, and the bicipital surface and deltoid crest are well formed. All these characteristics suggest a bird that had considerable powers of flight, and it is with the modern swifts that the closest match occurs.

HABITAT The successful early evolution of the swifts probably resulted from the increase in flowering plants and their insect pollinators. The latter provided an ample source of food for this insect predator.

Primapus lacki
Harrison & Walker;
London Clay;
Early Eocene; UK.

Typical length
6¾ in (17 cm)

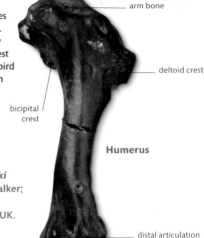

head of upper
arm bone

deltoid crest

bicipital
crest

Humerus

distal articulation

| Range: Early Eocene | Distribution: Europe | Occurrence: |

| Group: COLUMBIFORMES | Subgroup: RAPHIDAE | Informal name: Dodo |

Raphus

The enormous size of this flightless pigeon makes it easy to identify its fossilized remains. Notable bone features include a 6¾ in (17 cm) long head, with a distinct hooked bill; reduced wing elements; a keel-less breast bone; and short legs.

HABITAT *Raphus* was a ground dweller in wooded areas. It ate mainly fallen fruit and possibly grubs.
REMARK The extinction of the dodo during the 17th century was caused by the predation of humans and the animals they introduced.

Raphus cucullatus
(Linnaeus); Superficial
deposits; Quaternary;
Mauritius.

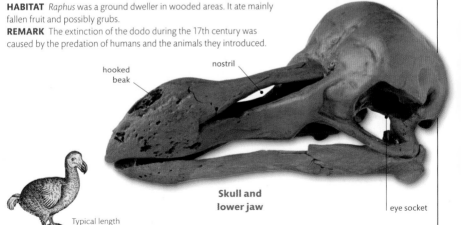

hooked
beak

nostril

**Skull and
lower jaw**

eye socket

Typical length
39 in (1 m)

| Range: Quaternary | Distribution: Mauritius | Occurrence: |

SYNAPSIDS

THE SYNAPSIDA IS A GROUP that includes all modern mammals and their ancestors, the latter erroneously referred to as mammal-like reptiles. They first appeared during the Carboniferous, more than 300 million years ago. Although the early examples were small and lizardlike, they were highly successful and adapted to a variety of habitats. By the Permian, they were the dominant land vertebrates, but they were severely depleted by the end Permian extinction event. Those that survived found it difficult to compete with the newly evolved and more agile archosaurs (crocodiles, dinosaurs, and their descendants) in the late Triassic. They did, however, give rise to the true mammals. The last surviving group of the synapsids, the tritylodonts, persisted into the early Cretaceous.

Group: EUPELYCOSAURIA	Subgroup: SPHENACODONTIDAE	Informal name: Predatory sail-back

Dimetrodon

The most spectacular feature of this genus was the huge dorsal sail, the result of the extreme elongation of the vertebral neural spines, which characteristically remain smooth along their length. The skull is deep, has daggerlike teeth, with the lower jaw articulation situated well below and behind the tooth row.

Dimetrodon loomisi Romer; Arroyo Formation; Early Permian; US.

HABITAT This relatively fast-moving, carnivorous reptile preyed upon less agile species in arid environments.
REMARK It was this group of sphenacodonts that acquired the adaptations that eventually developed fully in mammals.

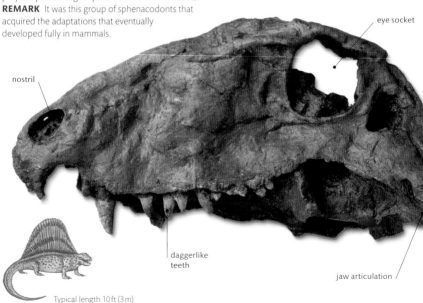

eye socket

nostril

daggerlike teeth

jaw articulation

Typical length 10 ft (3 m)

Range: Early Permian	Distribution: N. America	Occurrence:

| Group: EUPELYCOSAURIA | Subgroup: EDAPHOSAURIDAE | Informal name: Sailed herbivore |

Edaphosaurus

The overall body shape of this short-limbed, slow-moving quadruped was similar to that of a giant lizard, but the neural spines of the backbone extended upward to form a skin-covered sail. Individual spines have characteristic short transverse processes along their length. The short, broad skull contained both marginal and palatal teeth. The vertebral centra were strongly biconcave, a typical synapsid feature.

HABITAT This plant eater lived in arid areas.
REMARK The spectacular sail was probably a temperature control mechanism.

Vertebra

extended
neural spine

process of spine

Edaphosaurus sp.;
Wichita Formation;
Early Permian; US.

fragment
of palate,
with teeth

Typical length
10 ft (3 m)

| Range: Late Carboniferous–Early Permian | Distribution: Europe, N. America | Occurrence: |

| Group: THERAPSIDA | Subgroup: LYSTROSAURIDAE | Informal name: Lystrosaur |

Lystrosaurus

Skulls of lystrosaurs are characterized by their marked facial angles, the high placement of the nostrils, the possession of only two caninelike teeth in the upper jaw, and the massive but toothless lower jaw. They were heavy limbed, had short tails, and were quadrupedal in their gait.

HABITAT The angular skull shape suggests that this large herbivore was able to burrow to avoid environmental extremes.
REMARK *Lystrosaurus* survived the Permian-Triassic extinction event when few other animals did. They became abundant in the Early Triassic due to a lack of competition.

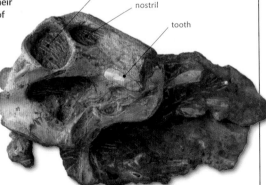

eye socket
nostril
tooth

Skull

Typical length
6½ ft (2 m)

Lystrosaurus murrayi
(Huxley); Karoo Formation;
Early Triassic; South Africa.

| Range: Late Permian–Early Triassic | Distribution: Worldwide | Occurrence: |

Group: THERAPSIDA	Subgroup: CYNOGNATHIDAE	Informal name: Cynodont

Cynognathus

This heavily built quadruped had a powerfully constructed skull, measuring some 16 in (40 cm) in length. The jaws were armed with paired, daggerlike canines and coarsely serrated cheek teeth. The lower jaw resembles those of large mammalian carnivores, with a large dorsally projected process toward its back, but the bones used in the skull articulation are more reptilian than mammalian.

HABITAT This ground-living carnivore hunted in arid country.

powerful muscle attachments

eye socket

large canine tooth

Cynognathus crateronotus Seeley; Karoo Formation; Early Triassic; South Africa.

Typical length 6½ ft (2 m)

Range: Early Triassic	Distribution: S. Africa, S. America	Occurrence:

Group: THERAPSIDA	Subgroup: TRITYLODONTIDAE	Informal name: Cynodont

Bienotherium

Many of the characteristic features of this highly advanced genus of therapsid can be seen in its skull. For example, there is a pair of greatly enlarged incisor teeth in the front of the upper and lower jaws, recalling those of present-day rodents, and the canines are absent, thus leaving a significant gap before reaching the single row of two to three cusped, squarish, molarlike, multiple-rooted cheek teeth. The latter, which are a mammalian feature, are unique to this family of reptiles. The cutting surfaces of the cheek teeth show three longitudinal ridges in the upper jaw and two in the lower.

Skull and lower jaw

eye socket

HABITAT *Bienotherium* was a gnawing plant eater.

Typical length 39 in (1 m)

paired incisors

Bienotherium yunnanense Young; Lower Lufeng Formation; Late Triassic; China.

Range: Early Jurassic	Distribution: Asia	Occurrence:

MAMMALS

MAMMALS ARE A VERY successful group. From terrestrial origins, they have colonized most of the habitable areas of the Earth's surface, the oceans, and the air. The main identifying features of mammals, such as the possession of hair, milk-producing mammary glands, and the details of their reproductive system, are rarely preserved as fossils. To the paleontologist, the most vital mammalian identification feature is the jaw articulation between the dentary bone (the only bone in the lower jaw of mammals) and the squamosal bone in the skull. The quadrate and articular bones, forming the articulation in other vertebrates, became associated with mammalian hearing and survive today as the incus and malleus of the middle ear.

| Group: MULTITUBERCULATA | Subgroup: TAENIOLABIDIDAE | Informal name: Multituberculate |

Taeniolabis

The multituberculates are an extinct order of primarily small, rodentlike animals characterized by the possession of highly distinctive teeth. One of the lower cheek teeth in each jaw was often enlarged to form a massive grinding tooth. Two-thirds of the mammals found in the Mongolian Cretaceous were multituberculates.

HABITAT This small rodent was a tropical forest dweller.

Taeniolabis taoensis (Cope); Early Paleocene; US.

enlarged grinding tooth

Lower jaw

mandibular bone

Typical length 24 in (60 cm)

| Range: Early Paleocene | Distribution: N. America | Occurrence: |

| Group: DRYOLESTIDA | Subgroup: DRYOLESTIDAE | Informal name: Dryolestid |

Amblotherium

Amblotherium was a small, insectivorous mammal, known mainly from isolated jaws. The genus was characterized by having a large number of teeth behind the canines, up to 12 per jaw. The individual teeth are broad and short, resembling those of the modern Madagascan tenrec, although the tenrec is a much more derived mammal.

HABITAT *Amblotherium* was a land dweller.

short, pointed insectivorous teeth

Amblotherium pusillum Owen; Purbeck Beds; Late Jurassic; UK.

Typical length 10 in (25 cm)

ramus of mandible

Lower jaw

| Range: Late Jurassic | Distribution: Europe, N. America | Occurrence: |

Group: DIDELPHIMORPHIA	Subgroup: DIDELPHIDAE	Informal name: Opossum

Didelphis

Opossums are mouse- to cat-sized marsupials with up to 50 teeth, a long snout, and small eyes. The pouch (or marsupium) is variably developed. Opossums have hands and feet well adapted for grasping; there are usually five digits on each foot, with the big toe acting as an opposable digit. While most species are climbers, some are not, and one is aquatic. Many species have prehensile tails.

HABITAT Opossums are arboreal, feeding on a wide variety of animal and vegetable matter.

Skull

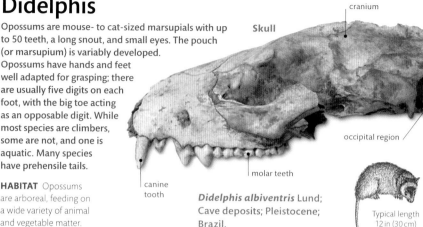

cranium

occipital region

molar teeth

canine tooth

Didelphis albiventris Lund; Cave deposits; Pleistocene; Brazil.

Typical length 12 in (30 cm)

Range: Pleistocene–Recent	Distribution: N. & S. America	Occurrence:

Group: DIPROTODONTIA	Subgroup: THYLACOLEONIDAE	Informal name: Marsupial lion

Thylacoleo

Thylacoleo was one of the more remarkable of the extinct members of the Australian marsupials and the largest of the indigenous carnivores. Lionlike in many respects, this animal was most closely related to the living phalangers or cuscuses, a group that now consists of small to medium-sized, arboreal herbivores. The wide, short-faced skull possesses large, paired front incisors, which seem to have fulfilled the functions of true carnivore canines, and massive, long, blade-like, shearing carnassial teeth for cutting up animal tissue. The braincase is small.

HABITAT This land mammal preyed on a wide variety of animals, many of which are now extinct, including *Diprotodon* (opposite), which was the largest of the native herbivores.

sagittal crest

Right profile of skull

Thylacoleo carnifex (Owen); River deposits; Pleistocene; Australia.

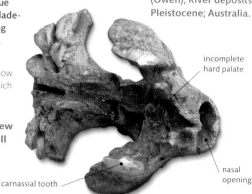

incomplete hard palate

nasal opening

Base view of skull

carnassial tooth

Typical length 5 ft (1.5 m)

Range: Pliocene–Pleistocene	Distribution: Australia	Occurrence:

| Group: DIRPROTODONTIA | Subgroup: MACROPODIDAE | Informal name: Kangaroo |

Procoptodon

This large kangaroo, now extinct, had notably heavy jaws and a relatively short face. Living kangaroos have smaller forelimbs than hindlimbs, being adapted for a primarily bipedal gait. The hindfoot was elongate and narrow, with unequal development of the digits. The cheek teeth had two dominant transverse shearing ridges, with a longitudinal connecting ridge. The paired lower incisors protruded forward. In each jaw, there was only one premolar, a shearing tooth, and this was followed by four molars, which erupted over a long period, moving forward in the jaw through the animal's life.

Procoptodon goliah
Owen; River deposits; Pleistocene; Australia.

HABITAT *Procoptodon* was adapted for browsing.

Typical length 10ft (3m)

Right lower jaw

empty tooth sockets

molar teeth

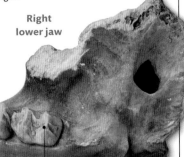

| Range: Pleistocene | Distribution: Australia | Occurrence: |

| Group: DIPROTODONTIA | Subgroup: DIPROTODONTIDAE | Informal name: Diprotodon |

Diprotodon

Diprotodon, the largest of the extinct giant marsupials, resembled a rhinoceros. It had huge, rodentlike incisor teeth, and cheek teeth each with two prominent transverse ridges, which wore down to form a shearing surface. The jaws were particularly thick and heavy. Skulls of *Diprotodon* are disproportionately massive, with a high nasal opening, large facial area, and small braincase.

Diprotodon australis
Owen; River deposits; Pleistocene; Australia.

HABITAT *Diprotodon* was a forest dweller, probably browsing on low-growing trees and shrubs.

Left lower jaw

two shearing facets on each cheek tooth

huge lower incisor

massive jaw bone

Typical length 10ft (3m)

| Range: Pleistocene | Distribution: Australia | Occurrence: |

Group: LIPOTYPHLA	Subgroup: TALPIDAE	Informal name: Russian desman

Desmana

Desmana is an aquatic insectivore, still found living in parts of Russia, Ukraine, and Kazakhstan. The short legs each bear five digits, with deeply grooved terminal phalanges. The bones of the forelimb are massive, with large muscle attachments. A short, thick clavicle is present, anchoring the limb to a prominent sternum. The dentition is typical of the insectivores, the molar teeth bearing many sharp, pointed cusps for dealing with an insect diet.

HABITAT *Desmana* was a common member of European lake and stream fauna in the Middle Pleistocene, making nesting chambers in muddy banks. It became extinct in western Europe 250,000 years ago.

large mandibular processes

Lower jaw

pointed cusps

Typical length 8 in (20 cm)

strong jaw

Desmana moschata (Linnaeus); West Runton Freshwater beds; Middle Pleistocene; UK.

Range: Pleistocene–Recent	Distribution: Europe, Asia	Occurrence:

Group: MICROCHIROPTERAMORPHA	Subgroup: ARCHEONYCTERIDAE	Informal name: Bat

Palaeochiropteryx

Entire skeleton

elongated digits

This bat had a complex ear region for echolocation and the characteristic wing structure of living bats, with the forelimbs modified into a long framework supporting the wing membranes. The first digit was used as a functional, grasping, clawed thumb, and the second digit was also clawed, like that of modern fruit bats. The third digit was the longest, extending to the wing tip. The membrane also extended between the fifth digit and the hindlimb and between the hindlimb and the tail.

HABITAT This bat lived off moths and flew close to the ground.

hindlimb bones

Typical length 2¾ in (7 cm)

Palaeochiropteryx tupaiodon Revilliod; Messel Formation; Eocene; Germany.

Range: Eocene	Distribution: Europe	Occurrence:

Group: PRIMATES	Subgroup: CERCOPITHECIDAE	Informal name: Macaque

Macaca

The macaques are a widespread group of medium-sized monkeys living in highly structured social groups. Large males can weigh up to 29 lb (13 kg). Their teeth are rather like smaller versions of their human equivalent, but the individual cusps are more prominent and more clearly defined.

HABITAT Macaques live in temperate woodland on a variety of plant and animal foods.
REMARK Never particularly common as fossils, they are Europe's only native Pleistocene primate, other than humans.

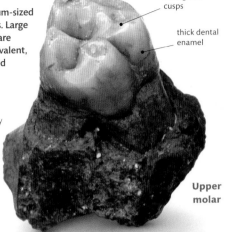

strong, low cusps

thick dental enamel

Upper molar

Typical length 28 in (70 cm)

Macaca pliocaena (Owen); Thames Terrace deposits; Middle Pleistocene; UK.

Range: Pleistocene–Recent	Distribution: Europe, N. Africa, N. Asia	Occurrence:

Group: XENARTHRA	Subgroup: GLYPTODONTIDAE	Informal name: Glyptodon

Glyptodon

The body of this armadillolike herbivore is almost completely clad in a bony armor coat, which forms a head shield, a domed carapace made up of hundreds of fused hexagonal scutes, and a dermal sheath to the tail. The forefeet bear large and powerful claws. *Glyptodon* has a short face, a very deep lower jaw, and massive zygomatic arches. The continuously erupting teeth are set very deeply into the jaw and lack an enamel coating.

HABITAT This giant herbivore lived in the pampas region of South America.

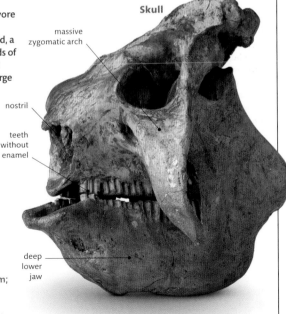

Skull

massive zygomatic arch

nostril

teeth without enamel

deep lower jaw

Typical length 6½ ft (2 m)

Glyptodon reticulatus Owen; Alluvium; Pleistocene; Argentina.

Range: Pleistocene–Recent	Distribution: S. America	Occurrence:

Group: XENARTHRA	Subgroup: MEGATHERIIDAE	Informal name: Giant ground sloth

Megatherium

Claw-bearing phalange

This giant sloth, now extinct, had a massive but short and high skull, with a high nasal opening and very deep jaws. It had very thick skin, armored throughout with small ossicles of bone, and huge, powerful claws on all four feet. Its very simple teeth, lacking in dental enamel, consisted of a battery of continuously growing square columns, deeply set in each jaw.

HABITAT This giant sloth lived on a wide range of vegetation.

channels for blood vessels

groove from which outer part of claw grew

deep articular notch

Typical length
20 ft (6 m)

Megatherium americanum
Cuvier; Alluvium;
Pleistocene; Argentina.

Range: Pliocene–Pleistocene	Distribution: N. & S. America	Occurrence:

Group: RODENTIA	Subgroup: GLIRIDAE	Informal name: Giant dormouse

Leithia

Underside of skull

Dormice are a primitive group of rodents, with very low-crowned grinding teeth which suit their mixed plant-food diet. Their teeth have biting surfaces divided into transverse ridges, which grind the food as if between two files. In common with many other herbivores, there is a long gap (or diastema) between the incisor teeth and the cheek teeth, dividing the mouth into two distinct functional regions, one dealing with procurement of food and the other with mastication.

HABITAT Probably nocturnal, this giant dormouse lived in woodland and dense scrub. It is likely to have hibernated in caves and holes in the ground.

diastema (gap between incisors and cheek teeth)

cheek teeth

transverse enamel ridges

Typical length
20 in (50 cm)

Leithia melitensis
Falconer; Cave breccia;
Pleistocene; Malta.

Range: Pleistocene	Distribution: Malta	Occurrence:

Group: RODENTIA	Subgroup: ARVICOLIDAE	Informal name: Water vole

Arvicola

Like many other voles, this water vole can easily be identified by its teeth. It has an enlarged, platelike zygomatic arch. Its high-crowned cheek teeth are made up of triangular prisms. The enamel pattern of the largest lower molar identifies the genus. The skull is slightly flattened, in keeping with the predominantly burrowing habits of many family members.

HABITAT Voles form extensive shallow burrows for nesting and feeding.

back of skull missing

Part of skull

high-crowned cheek teeth

incisor

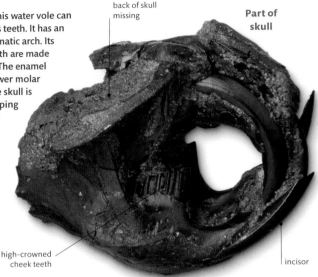

Typical length 2¾ in (7 cm)

Arvicola cantiana
(Hinton); Estuarine silts;
Middle Pleistocene; UK.

Range: Pleistocene–Recent	Distribution: Europe, Asia	Occurrence:

Group: CETARTIODACTYLA	Subgroup: BASILOSAURIDAE	Informal name: Whale

Basilosaurus

The basilosaurid whale had a long body, little or no hindlimbs, and forelimbs modified into paddles. The skull underwent relatively little modification, other than elongation of the muzzle and the development of specialized teeth. The nasal region was little modified, and the posterior part of the skull closely resembled contemporary terrestrial carnivores, with large, zygomatic arches and a marked sagittal crest.

HABITAT Basilosaurids lived in warm, shallow seas.

Basilosaurus sp.;
Mokattam Formation;
Eocene; Egypt.

olfactory lobe

cerebral hemisphere

Natural cast of brain

cerebellum

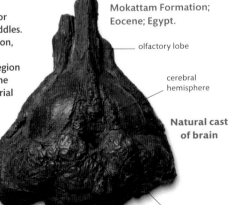

Typical length 53 ft (16 m)

Range: Eocene	Distribution: Worldwide	Occurrence:

| Group: CETARTIODACTYLA | Subgroup: BALAENIDAE | Informal name: Right whale |

Balaena

Balaena has developed a highly specialized filtration system to replace its dentition. The filter is composed of baleen plates—like the bristles of a huge brush—and is used to filter out small sea creatures, such as krill. The nasal opening has migrated to a position high on the forehead to form a closable blowhole. There is no outward trace of a rear limb, and the front limb is modified into a huge paddle, articulated on a very short humerus. Whales have distinctive tympanic (ear) bones, quite commonly found as fossils. In baleen whales, these actually resemble a large ear.

HABITAT *Balaena* lives in temperate to Arctic seas.

Ear bone (Tympanic)

Balaena primigenia
Van Beneden; Red Crag; Pliocene; UK.

polished outer surface

preservation typical of the Red Crag

Typical length 65 ft (20 m)

| Range: Pliocene–Recent | Distribution: Worldwide | Occurrence: |

| Group: HYAENODONTA | Subgroup: HYAENODONTIDAE | Informal name: Creodont |

Hyaenodon

This large, specialized carnivore was terrestrial, walking on its toes (digitigrade). It had well-developed, flesh-piercing canine teeth, and the upper first and second molars, along with the lower second and third molars, developed into carnassial (meat-slicing) teeth. In modern carnivores, the carnassials are the last upper premolar and first lower molar.

HABITAT *Hyaenodon* was a land-dwelling carnivore.

Hyaenodon horridus
Leidy; Fluviatile deposits; Oligocene; US.

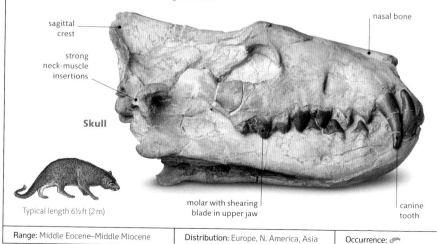

nasal bone

sagittal crest

strong neck-muscle insertions

Skull

Typical length 6½ ft (2 m)

molar with shearing blade in upper jaw

canine tooth

| Range: Middle Eocene–Middle Miocene | Distribution: Europe, N. America, Asia | Occurrence: |

| Group: CARNIVORA | Subgroup: URSIDAE | Informal name: Bear |

Ursus

Ursus refers to the cave, the brown, and the grizzly bear. Fossil remains are usually of the cave bear. Its characteristics include a skull with a high forehead, loss of the anterior premolars, and low-crowned cheek teeth with many tiny cusps on the crushing surfaces.

HABITAT They lived in temperate woodland. In winter, they hibernated in caves.
REMARK Many European caves have been found to contain extraordinary quantities of cave bear remains.

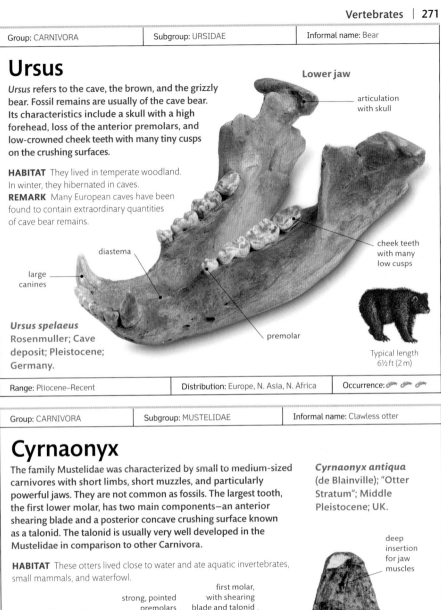

Lower jaw

articulation with skull

cheek teeth with many low cusps

diastema

large canines

Ursus spelaeus Rosenmuller; Cave deposit; Pleistocene; Germany.

premolar

Typical length 6½ ft (2 m)

| Range: Pliocene–Recent | Distribution: Europe, N. Asia, N. Africa | Occurrence: |

| Group: CARNIVORA | Subgroup: MUSTELIDAE | Informal name: Clawless otter |

Cyrnaonyx

The family Mustelidae was characterized by small to medium-sized carnivores with short limbs, short muzzles, and particularly powerful jaws. They are not common as fossils. The largest tooth, the first lower molar, has two main components—an anterior shearing blade and a posterior concave crushing surface known as a talonid. The talonid is usually very well developed in the Mustelidae in comparison to other Carnivora.

Cyrnaonyx antiqua (de Blainville); "Otter Stratum"; Middle Pleistocene; UK.

HABITAT These otters lived close to water and ate aquatic invertebrates, small mammals, and waterfowl.

deep insertion for jaw muscles

first molar, with shearing blade and talonid

strong, pointed premolars

large canine tooth

Typical length 5 ft (1.5 m)

Lower jaw

| Range: Pleistocene | Distribution: Europe, Asia | Occurrence: |

Group: CARNIVORA	Subgroup: FELIDAE	Informal name: Big cat

Panthera

The first lower molar of this large cat has a highly efficient flesh-shearing notch that works against the blade of the corresponding last upper premolar. The canine tooth is large and conical, with two prominent parallel grooves in its enamel surface, typical of the family.

Panthera leo
(Linnaeus); Fluviatile deposits; Middle Pleistocene; UK.

HABITAT *Panthera* was an open-country predator.
REMARK Fossil lions are closely associated with fossil horses, and as the horses of the Middle Pleistocene were large, so, too, were the lions.

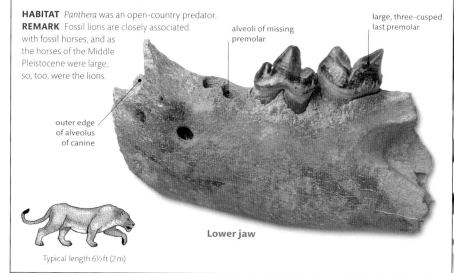

alveoli of missing premolar

large, three-cusped last premolar

outer edge of alveolus of canine

Lower jaw

Typical length 6½ ft (2 m)

Range: Miocene–Recent	Distribution: Northern hemisphere	Occurrence:

Group: PHENACODONTA	Subgroup: PHENACODONTIDAE	Informal name: Early ungulate

Phenacodus

low, blunt cusps

molar tooth

This genus is known from the entire skeleton. The size ranges from that of a medium to a large dog. Each foot bore five toes, each with a small hoof. The teeth had low, blunt cusps, suitable for a diet consisting mainly of fruit. The family is considered close to the stem of the Perissodactyla (horses, rhinoceroses, and tapirs), the Proboscidea (elephants), and the Sirenia (sea cows).

HABITAT *Phenacodus* was a ground dweller that fed mainly on fallen fruit.

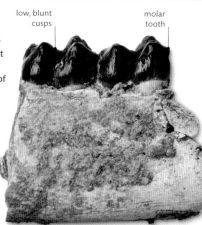

Phenacodus vortmani (Cope); Willwood Formation; Early Eocene; US.

Typical length 36 in (90 cm)

Fragment of lower jaw

Range: Late Paleocene–Middle Eocene	Distribution: N. America, Europe	Occurrence:

| Group: NOTOUNGULATA | Subgroup: TOXODONTIDAE | Informal name: Notoungulate |

Toxodon

This was a large grazing animal the size and build of a rhinoceros, with robust limb bones and a short neck. The feet bore three hoofed toes.

HABITAT The incisors and cheek teeth were continuously growing, suggesting an abrasive diet, such as grass.

REMARK The notoungulates, like the litopterns (below), evolved in isolation in South America during the Tertiary. *Toxodon*, like *Macrauchenia*, was the last survivor of its order.

Toxodon platensis Owen; Fluviatile gravels; Pleistocene; Argentina.

eye socket

short nasal bone

Skull

Typical length 10 ft (3 m)

continuously growing cheek teeth

alveoli for chisel-shaped incisors

| Range: Pleistocene | Distribution: S. America | Occurrence: |

| Group: LITOPTERNA | Subgroup: MACRAUCHENIIDAE | Informal name: Litoptern |

Macrauchenia

This large, camel-like mammal had a long neck and three-toed feet. Its external nasal opening, instead of being at the front of the skull, was in the skull roof between the eyes. This has been interpreted as denoting either an aquatic mode of life or the presence of a trunk.

registration number

HABITAT *Macrauchenia* is thought to have browsed on the edges of lowland forests.

REMARK This specimen was collected by Charles Darwin in 1834 on the famous voyage of the *Beagle*. It still bears the registration number given by the Royal College of Surgeons.

Macrauchenia patachonica Owen; Fluviatile gravels; Pleistocene; Argentina.

long toe bone

Bones of right forefoot

Typical length 10 ft (3 m)

| Range: Pleistocene | Distribution: S. America | Occurrence: |

| Group: PROBOSCIDEA | Subgroup: PHIOMIIDAE | Informal name: Phiomiid |

Phiomia

Phiomia had short, protruding, flattened lower tusks (incisors) and longer, downcurved upper tusks. The cheek teeth are low, with conical cusps arranged in three pairs (bunodont). A diastema is present between the tusks and cheek teeth. The retracted position of the nasal opening suggests there was a short trunk.

HABITAT *Phiomia* probably fed on low-growing shrubs.

REMARK It is likely that the upper tusks were used for defensive purposes.

flattened lower tusks

diastema

articulates with upper jaw

cusps in three rows

low-crowned cheek teeth

Phiomia serridens
Andrews & Beadnell;
Jebel Qatrani Formation;
Oligocene; Egypt.

Near-complete, paired mandibles

Typical shoulder height
8 ft (2.4 m)

| Range: Oligocene | Distribution: N. Africa | Occurrence: |

| Group: PROBOSCIDEA | Subgroup: GOMPHOTHERIIDAE | Informal name: Gomphothere |

Tetralophodon

Tetralophodon represents the stem group for the true elephants. The head was long and often lacked mandibular tusks, although those from the upper jaw were often quite long. The cheek teeth bear a series of paired, conical, "bunodont" cusps, similar to the mastodon's.

Tetralophodon longirostris Kaup;
Middle Miocene;
France.

paired, conical cusps

HABITAT *Tetralophodon* is thought to have browsed on vegetation beyond the reach of other herbivores.

root

Typical shoulder height
8 ft (2.5 m)

Cheek tooth

| Range: Miocene–Pliocene | Distribution: Europe, Asia, N. America | Occurrence: |

Group: PROBOSCIDEA	Subgroup: DEINOTHERIIDAE	Informal name: Deinothere

Deinotherium

The deinotheres are an extinct group of elephantlike proboscideans, lacking tusks in the upper jaw but possessing a large pair of downturned tusks in the lower jaw. The cheek teeth are characterized by having two or three simple transverse ridges (lophs). These were used to shear plant material, as opposed to the crushing action that was more common in most other more primitive proboscideans.

HABITAT *Deinotherium* was probably a forest dweller. Wear patterns suggest that the downturned tusks were used for digging roots or stripping tree bark.

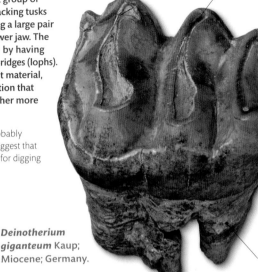

Cheek tooth

transverse ridges, with enamel worn through to dentine

tooth root

Deinotherium giganteum Kaup; Miocene; Germany.

Typical shoulder height
13 ft (4 m)

Range: Miocene–Pleistocene	Distribution: Europe, Asia, Africa	Occurrence:

Group: PROBOSCIDEA	Subgroup: MAMMUTIDAE	Informal name: Mastodon

Mammut

A contemporary of the mammoths, the North American mastodon was a member of a morphologically more basal group of proboscideans, with low-crowned or "bunodont" cheek teeth covered in thick enamel and with large, rounded, crushing cusps.

HABITAT Mastodons were forest dwellers.
REMARK Their extinction in the late Pleistocene is thought to be due to a combination of dietary inflexibility, climate-induced habitat loss, and hunting.

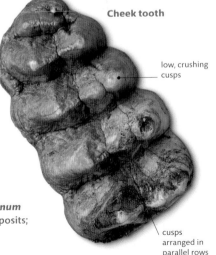

Cheek tooth

low, crushing cusps

cusps arranged in parallel rows

Mammut americanum (Cuvier); Spring deposits; Pleistocene; US.

Typical shoulder height
8 ft (2.5 m)

Range: Miocene–Pleistocene	Distribution: N. America	Occurrence:

| Group: PROBOSCIDEA | Subgroup: ELEPHANTIDAE | Informal name: Mammoth |

Mammuthus

At about the same size as an Asian elephant, mammoths must have proved a formidable prey for early human hunters. Mammoths first appeared in the African Pliocene and rapidly spread to Europe and Asia. Their evolution can most readily be traced through their tooth structure. The teeth of a mammoth consists of a series of plates composed of enamel surrounding a dentine core. Each tooth erupted from the back of the jaw and slowly moved forward as it wore, to be replaced by another tooth from behind. The thickness and number of tooth plates are important identification criteria.

HABITAT Mammoths are thought to have grazed on grasses and low shrubs. Whole frozen carcasses have been found in the Siberian Arctic. The carcasses confirm the accuracy of drawings of long-haired mammoths in European caves.

Mammuthus primigenius Blumenbach; Permafrost deposits; Pleistocene; Siberia.

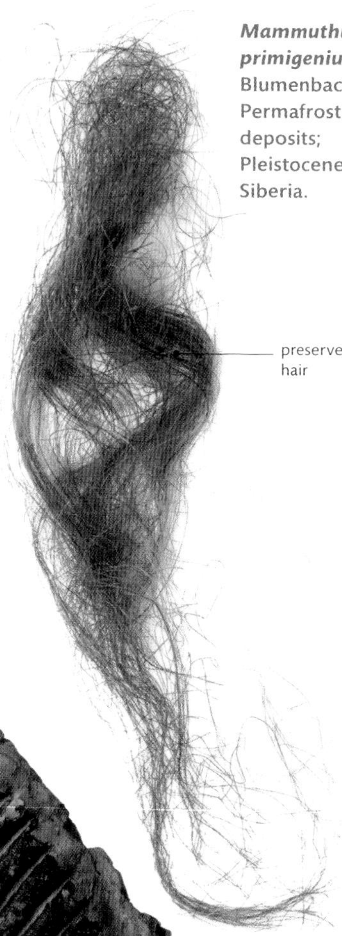

preserved hair

Hair

Mammuthus primigenius Blumenbach; Glacial gravels; Pleistocene; UK.

matrix of dental cement

series of enamel plates

Upper cheek tooth

Typical shoulder height 10 ft (3 m)

flat chewing surface

| Range: Pliocene–Early Holocene | Distribution: Europe | Occurrence: |

| Group: HYRACOIDEA | Subgroup: PLIOHYRACIDAE | Informal name: Giant hyrax |

Titanohyrax

Resembling a very large guinea pig, *Titanohyrax* was the largest known species of hyrax, about the size of a modern tapir. The incisor teeth were enlarged for nibbling and the cheek teeth were rhinoceroslike and quadrate, with a thick enamel coat and prominent lophs.

HABITAT It is probable that *Titanohyrax* browsed on scrub vegetation.

Typical length 6½ft (2 m)

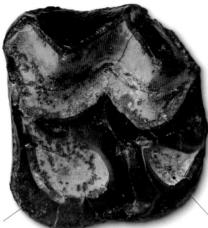

Titanohyrax ultimus Matsumoto; Jebel Qatrani Formation; Oligocene; Egypt.

Upper cheek tooth

enamel

prominent dentine lophs

| Range: Eocene–Early Oligocene | Distribution: N. Africa | Occurrence: |

| Group: PERISSODACTYLA | Subgroup: EQUIDAE | Informal name: Early horse |

Xenicohippus

This genus of early horse is related to the true horses, which evolved in North America, and the paleotheres, which evolved in isolation in Europe and became extinct in the Oligocene. *Xenicohippus* had four toes on the forefeet and three toes on the hind feet, each of which bore a small hoof. It was adapted for running, but its gait was less erect than that of a modern horse. Its teeth were low-crowned with low cusps, which enabled it to live off a diet of both fruit and soft leaves.

HABITAT *Xenicohippus* is thought to have lived in a tropical forest environment. It was previously included in *Hyracotherium*.

Xenicohippus craspedotum Cope; Wind River Formation; Early Eocene; US.

Maxilla with cheek teeth

simple, low-crowned tooth

fragment of upper jaw

Typical shoulder height 16 in (40 cm)

| Range: Early Eocene | Distribution: Europe, N. America | Occurrence: |

Group: PERISSODACTYLA	Subgroup: EQUIDAE	Informal name: Horse

Equus

Among the best known of all the mammals, the horse is nearly extinct in the wild, but survives in domestication. A grazing, terrestrial herbivore, the horse has square, high-crowned cheek teeth with complex enamel patterns. The feet are heavily modified, being reduced to a single, elongate metapodial with a shore phalange terminating on a prominent hoof, thus adapting them for rapid forward movement on hard ground. The anatomy and physiology of the digestive system, with a large caecum and colon, enables horses to subsist upon high-fiber diets with a low protein content.

HABITAT Modern horses and their relatives have teeth and limbs that are adapted for a grazing, plains-dwelling life.

REMARK The genus *Equus* first appeared in North America in the Pliocene and rapidly spread to every continent except Australia and Antarctica. Cave paintings dating from the Paleolithic period indicate that the Late Pleistocene forms resembled the extant Przewalskii's horse, with a mane of short, stiff, upwardly pointing hair.

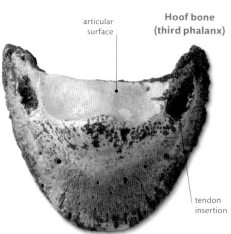

Hoof bone (third phalanx)

articular surface

tendon insertion

Equus ferus Boddaert; Cave earth; Late Pleistocene; UK.

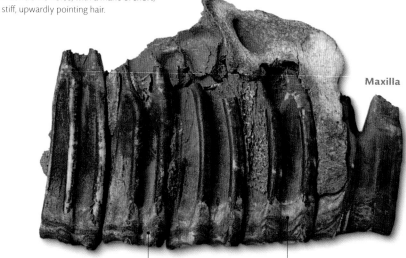

Maxilla

molars

premolars

Typical shoulder height 5 ft (1.5 m)

Range: Pleistocene–Recent	Distribution: Worldwide	Occurrence:

| Group: PERISSODACTYLA | Subgroup: PARACERATHERIIDAE | Informal name: Giant hornless rhino |

Paraceratherium

Together with the closely related *Indricotherium*, *Paraceratherium* was the largest land mammal ever to have lived. It could be described as a gigantic, long-legged, hornless rhinoceros. The absence of a horn has been deduced from the long, slender nasal bones, which are too weak to support a horn. Its teeth bear a similar pattern to those of the modern rhinoceroses.

HABITAT This land mammal was adapted for tree-top browsing.

Fragment of maxilla

dentine exposed by wear

cheek tooth

hard palate

Paraceratherium bugtiense
Cooper; Bugti Beds; Oligocene; Pakistan.

Typical shoulder height 16½ ft (5 m)

| Range: Oligocene | Distribution: Asia | Occurrence: |

| Group: PERISSODACTYLA | Subgroup: RHINOCEROTIDAE | Informal name: Woolly rhinoceros |

Coelodonta

The most striking features of the woolly rhinoceros were its prominent shoulder hump, shaggy coat, and two horns arranged in tandem. The cheek teeth are high-crowned, with thick, rugose enamel and a heavy covering of dental cement.

HABITAT The woolly rhinoceros was a grazer, feeding upon tundra grasses and low-growing shrubs.
REMARK European cave paintings show a similar animal, with two horns and a shaggy coat.

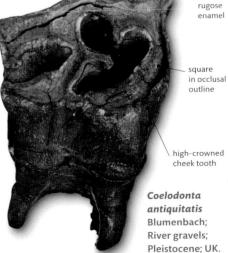

thick, rugose enamel

square in occlusal outline

high-crowned cheek tooth

Coelodonta antiquitatis
Blumenbach; River gravels; Pleistocene; UK.

Second upper molar

Typical length 13 ft (4 m)

| Range: Miocene–Pleistocene | Distribution: Europe, Asia | Occurrence: |

Group: CETARTIODACTYLA	Subgroup: GIRAFFIDAE	Informal name: Giraffe

Samotherium

This was a relatively short-necked giraffe with a pair of horns just behind the eyes. The muzzle was elongated, with a long diastema and a rounded end. The cheek teeth are high-crowned and adapted for sideways chewing, much like those of a sheep or cow.

Samotherium boissieri Major; Tuffs; Late; Miocene; Greece.

HABITAT The rounded muzzle is regarded as an adaptation to grazing, so it is likely that *Samotherium* inhabited open grasslands. In contrast, the modern giraffids (the giraffe and okapi) are browsers.

medium-crowned cheek teeth

sharp, crescentic ridges

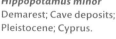

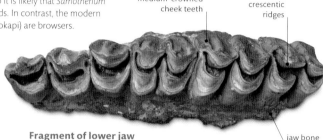

Typical length 10 ft (3 m)

Fragment of lower jaw

jaw bone

Range: Miocene	Distribution: Europe, Asia, Africa	Occurrence:

Group: CETARTIODACTYLA	Subgroup: HIPPOPOTAMIDAE	Informal name: Hippopotamus

Hippopotamus

Hippopotamuses are principally short-legged, stocky, aquatic herbivores. They have long, protruding lower incisors; huge canine teeth (tusks); a short diastema; and a battery of high-crowned cheek teeth. They can exceed 3 tons in weight, although the dwarf species illustrated here was the size of a small pig.

alveolus (socket) of tusk

HABITAT Because of their great bulk, hippopotamuses are cumbersome on land but are surprisingly agile in water, where they graze on grasses, reeds, and other aquatic plants.
REMARK The illustrated specimen is a dwarf species. Dwarf and giant forms are typical of island faunas.

Hippopotamus minor Demarest; Cave deposits; Pleistocene; Cyprus.

sockets of premolars

high-crowned molars

Typical length 39 in (1 m)

Incomplete skull

Range: Early Pliocene–Recent	Distribution: Europe, Asia, Africa	Occurrence:

Group: CETARTIODACTYLA	Subgroup: MERYCOIDODONTIDAE	Informal name: Oreodont

Merycoidodon

This genus, formerly known as *Oreodon*, is distantly related to the camels, although this was not reflected in its outward appearance. It was a short, stocky animal, the size of a sheep. It looked like a short-legged deer or horse but had four toes on each foot. It had medium-crowned, crescentic cheek teeth and no diastema.

HABITAT During the Oligocene, herds of *Merycoidodon* roamed the woodlands and prairies of North America, browsing on leaves.

Merycoidodon culbertsoni
Leidy; Brule Formation; Oligocene; US.

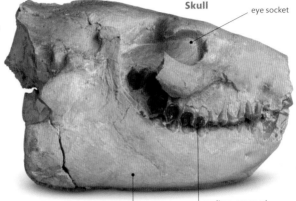

Skull

eye socket

medium-crowned, crescentic teeth

deep, robust mandible

Typical length 5 ft (1.5 m)

Range: Oligocene	Distribution: N. America	Occurrence:

Group: CETARTIODACTYLA	Subgroup: BOVIDAE	Informal name: Bison

Bison

Bison have true horns, comprising a core of bone covered by a horn sheath in a single, upright curve. These are found in both sexes. Mature bulls use their horns for both fighting and display. The lower incisor teeth are spatulate; the uppers are absent. The cheek teeth are high-crowned.

Bison priscus Bojanus; Floodplain terrace; Late Pleistocene; UK.

horn core

HABITAT Herds of bison have roamed temperate grasslands in North America and Europe from the Pleistocene to the present day.

Skull roof

frontal bone

eye socket

Typical length 10 ft (3 m)

Range: Pliocene–Recent	Distribution: Europe, N. America	Occurrence:

Group: PRIMATES	Subgroup: HOMINIDAE	Informal name: Gigantopithecine

Gigantopithecus

This was an enormous, gorillalike, ground-living ape, possibly the largest primate ever to have lived. Estimates of its weight are in excess of 600 lb (273 kg). This primate had robust canine teeth and huge cheek teeth, with thick enamel, high crowns (for a primate), and low cusps.

HABITAT The degree of wear on the teeth suggests that they were used heavily in mastication of a specific kind of plant food, possibly seeds.

REMARK The genus survived in China at least until 500,000 years ago.

Gigantopithecus blacki von Konigswald; Cave deposits; Early Pleistocene; China.

Typical height 6½ ft (2 m)

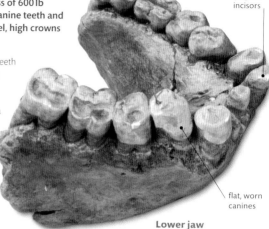

extensive occlusal surfaces

peglike incisors

flat, worn canines

Lower jaw

Range: Miocene–Pleistocene	Distribution: Asia	Occurrence:

Group: PRIMATES	Subgroup: HOMINIDAE	Informal name: Australopithecine

Paranthropus

The australopithecines were characterized by robust skulls with powerful lower jaws, and bipedal locomotion. Their canines were smaller than those of their presumed ancestors, but the cheek teeth were large, with thick enamel, suitable for a tough, herbivorous diet.

HABITAT They lived in open country and woodland environments in tropical and subtropical areas.

Paranthropus boisei (Leakey); Shungura Formation; Pliocene; Ethiopia.

Typical height 4½ ft (1.4 m)

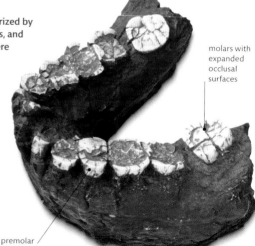

molars with expanded occlusal surfaces

premolar

Lower jaw

Range: Pliocene–Early Pleistocene	Distribution: Africa	Occurrence:

Group: PRIMATES	Subgroup: HOMINIDAE	Informal name: Early human

Homo habilis

Homo habilis had a larger and more rounded skull than any earlier hominid. The incisor teeth were relatively large; the premolars were small. The molars were narrow, with thick enamel.

HABITAT *Homo habilis* lived in the open savannah.

REMARK The dental features, combined with evidence of toolmaking, point to a possible dependence, for the first time in hominids, on meat-eating and hunting.

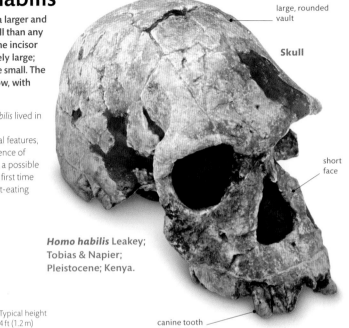

large, rounded vault

Skull

short face

Homo habilis Leakey; Tobias & Napier; Pleistocene; Kenya.

Typical height 4 ft (1.2 m)

canine tooth

Range: Late Pliocene–Early Pleistocene	Distribution: Africa	Occurrence:

Group: PRIMATES	Subgroup: HOMINIDAE	Informal name: Neanderthal

Homo neanderthalensis

The neanderthals were short and heavily muscled, with prominent brow ridges and a receding forehead and chin. The body shape probably resulted from a long process of climatic adaptation. They had a brain size similar to that of modern humans, were hunters, fashioned complex wood and bone tools, used fire, and buried their dead. They inhabited Europe and western Asia during the last ice age.

Homo neanderthalensis King; Cave earth; Pleistocene; France.

HABITAT Neanderthals lived in temperate and cold climates.

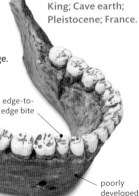

gap behind third molar

edge-to-edge bite

Lower jaw of young adult

Typical height 5¹⁄₄ ft (1.6 m)

poorly developed chin

Range: Late Pleistocene	Distribution: Europe, Asia	Occurrence:

| Group: PRIMATES | Subgroup: HOMINIDAE | Informal name: Modern human |

Homo sapiens

Modern humans are taller than the neanderthals, with more gracile bones, more prominent chins, and more domed foreheads. Their teeth are similar, though the incisors are less protruding.

HABITAT Although once restricted to warm climates, the acquisition of the skills to make clothing and construct shelters allowed modern humans to colonize more hostile environments. Cooperative hunting, armed with simple stone and wood weapons, allowed them to exploit prey that would otherwise have eluded them.

REMARK Modern humans manufactured a variety of tools with specific functions, as distinct from the multipurpose axes of their forbears. The bone, flint, and antler tools shown opposite are often found associated with fossil human remains, although not strictly speaking fossils themselves. Although of different ages and degrees of sophistication, they indicate a level of culture rather than a specific date.

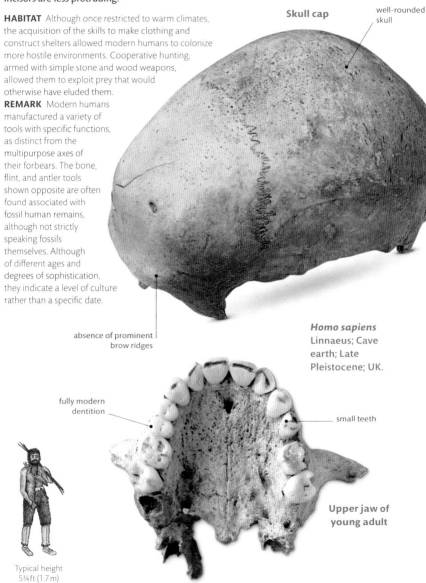

Skull cap

well-rounded skull

absence of prominent brow ridges

Homo sapiens
Linnaeus; Cave earth; Late Pleistocene; UK.

fully modern dentition

small teeth

Upper jaw of young adult

Typical height
5¾ ft (1.7 m)

| Range: Late Pleistocene–Recent | Distribution: Worldwide | Occurrence: |

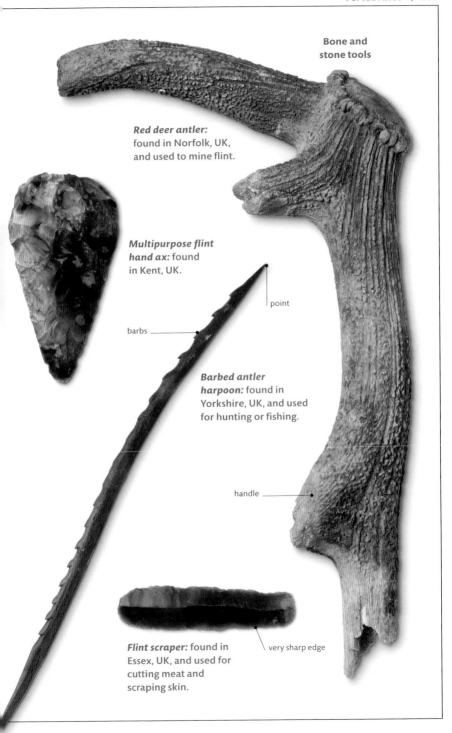

Bone and stone tools

Red deer antler: found in Norfolk, UK, and used to mine flint.

Multipurpose flint hand ax: found in Kent, UK.

point

barbs

Barbed antler harpoon: found in Yorkshire, UK, and used for hunting or fishing.

handle

Flint scraper: found in Essex, UK, and used for cutting meat and scraping skin.

very sharp edge

ALGAE

THE ALGAE HAVE A fossil record that extends from the Precambrian to the present day. Some algae are important age indicators and are used extensively in the oil industry. They present a wide range of forms, from simple unicells to complex multicellular plants. Fossil remains are usually limited to those that produce structures impregnated with silica or forms of calcium carbonate or that developed tough-walled cysts. As algae released oxygen into the atmosphere, they were responsible for a critical change in atmospheric composition during the Precambrian about 2.4 billion years ago. This is referred to as "The Great Oxidation Event," which set the stage for complex multicellular life in the oceans.

Group: CYANOPHYTA	Subgroup: SPONGIOSTROMATA	Informal name: Stromatolites

Collenia

Formed by cyanobacteria, this stromatolite has a conical to cylindrical structure composed of limestone and/or silica. This structure is always layered, sometimes with alternating light and dark bands. The minute algal threads of which it is composed can be seen under the microscope. These threads bound together detrital sands and muds in a bed of lime produced by the algae.

HABITAT Like living stromatolitic algae, *Collenia* inhabited intertidal zones.

Collenia sp.;
Stromatolitic
Limestone;
Precambrian; US.

laminated structure

limestone

bonded detrital and organic materials

Typical height
10 ft (3 m)

Range: Precambrian–Cambrian	Distribution: Worldwide	Occurrence: ✿

| Group: DASYCLADALES | Subgroup: DASYCLADACEAE | Informal name: Green algae |

Mastopora

The genus is characterized by clusters of thalli originating from a central axis. The resulting globular structure was covered by a calcified mucillage, which helped it to be preserved as a fossil. The reticulum was formed by primary branches radiating from a central axis; each branch terminated bluntly and secreted a protective limestone covering. The fossils have a characteristic hexagonal honeycomb pattern on the surface. Each hexagon has a raised border with a depressed center, imparting a rough surface to the fossil.

HABITAT *Mastopora* are usually associated with fossil coral reefs, brachiopods, and bryozoans.

REMARK This genus of plants was previously and inaccurately classified as an animal (protozoan or sponge).

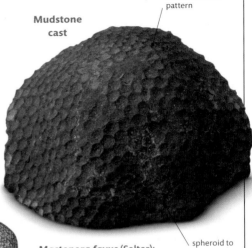

Mudstone cast

honeycomblike pattern

spheroid to ovoid body

Mastopora favus (Salter); Shelly limestone; Silurian; UK.

Typical height 3 in (8 cm)

| Range: Ordovician–Silurian | Distribution: Worldwide | Occurrence: ✿ |

| Group: RECEPTACULITALES | Subgroup: RECEPTACULITACEAE | Informal name: Receptaculitid |

Ischadites

The receptaculitids are globe-shaped fossils composed of a series of elements with rhomboidal ends, originating from the center. The wall of the head consists of spirally arranged, diamond-shaped plates. The thallus wall itself was calcified rather than having a mucillage covering, as in other calcareous algae.

HABITAT In life, the axis anchored the plant to rocks or reef corals.

globose shape

depressed top

spiral-patterned ornament

Calcified thallus

Ischadites barrandei (Hinde); Limestone; Silurian; Czech Republic.

Typical height 2 in (5 cm)

| Range: Ordovician–Silurian | Distribution: Worldwide | Occurrence: ✿ ✿ |

Group: PHAEOPHYTA	Subgroup: Unclassified	Informal name: Brown alga

Bythotrephis

These soft-bodied algae were small to moderately large plants of simple, purely vegetative structure, with "Y"-fork branches. The fossils are present as impressions or carbonaceous films on the rock surfaces. There is often no detailed preservation, although linear striations, which indicate the remains of tube structures within the body of the plant, may be present on the branches.

HABITAT *Bythotrephis* can be found in shallow (or shoreline) to deep-water sediments.
REMARK Care is needed when identifying these fossils, as they are similar to burrow systems of small marine invertebrates. The presence of carbonaceous or similar minerals is the only certain guide to a correct identification.

Typical height
12 in (30 cm)

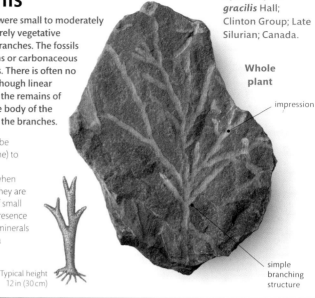

Bythotrephis gracilis Hall; Clinton Group; Late Silurian; Canada.

Whole plant

impression

simple branching structure

Range: Silurian–Miocene	Distribution: Worldwide	Occurrence: ✿ ✿

Group: CHLOROPHYTA	Subgroup: Unclassified	Informal name: Green alga

Parka

The liverwortlike thallus or plant body of *Parka* had a thick outer covering or cuticle. The plant body itself sometimes had a simple "Y" branch and was only tens of cells thick. It nearly always bore distinct rounded bodies or spore capsules, which contained numerous small spores.

HABITAT The protective outer covering to both plant and spores may indicate that *Parka* was, in fact, a land plant.

Fragment of thallus

Parka decipiens
Flemming; Carmyllie beds; Early Devonian; UK.

Typical diameter
1½ in (4 cm)

spore capsules   mudstone

carbonaceous layer

Range: Late Silurian–Early Devonian	Distribution: Europe, N. America	Occurrence: ✿ ✿

EARLY LAND PLANTS

THE FIRST INDISPUTABLE land plants are characterized by a mechanical supporting tissue, which provided rigidity; a perforated cuticle, which allowed the plant to "breathe"; and spores resistant to desiccation, due to a tough spore coat, which helped them survive and germinate on land. Spore-bearing capsules were borne either at the ends of thin branches, singly or in bunches, or along the sides. These plants ranged from small, scrambling types to spiny-stemmed varieties 3¼ ft (1 m) tall.

Group: RHYNIOPHYTES	Subgroup: COOKSONIOIDEA	Informal name: Cooksonia

Cooksonia

These small plants were characterized by "Y" branches. The simplest consisted of one or two branches, but some had five to six levels of branching. Typically, each branch had a spore capsule at the end, which was oval or kidney-shaped and contained spores with resistant coats.

HABITAT The earliest known land plant, its remains are usually associated with shallow marine or fluvio-deltaic deposits.

Typical height 3 in (7.5 cm)

shelly mudstone

spore capsule

"Y" branch

Cooksonia hemisphaerica Lang; Shelly mudstone; Early Devonian; UK.

Range: Late Silurian–Devonian	Distribution: Worldwide	Occurrence: ☼

Group: TRACHEOPHYTES	Subgroup: LYCOPHYTES	Informal name: Zoscerophyll

Zosterophyllum

Zosterophyllum is a genus of plants that had spore capsules placed near the end of the stem, but always on the side, never at the extremity. The spore capsules were oval shaped and on short stalks. The branching was a mixture of "Y" type (these were less pronounced) and "H" type.

HABITAT *Zosterophyllum* grew mainly on the margins of lakes.

REMARK It is probable that plants like these gave rise to the giant lycopods of the coal swamps, such as *Lepidodendron* (see p.294).

Typical height 10 in (25 cm)

spore capsules on stem side

Zosterophyllum llanoveranum Croft & Lang; Senni beds; Early Devonian; UK.

mudstone

stems

Range: Late Silurian–Middle Devonian	Distribution: Worldwide	Occurrence: ☼

HEPATOPHYTES

HEPATOPHYTES OR LIVERWORTS are small plants with flat (thalloid) or leafy bodies. They reproduce by spores contained in capsules borne on wiry stems. Today, these plants are found in damp environments worldwide.

Group: MARCHANTIALES	Subgroup: MARCHANTIACEAE	Informal name: Liverwort

Hexagonocaulon

This small creeping plant had a noticeable "Y"-branch pattern every 1 or 2 centimeters. The plant body was flat and broad. Small rootlike structures were present on the underside in life. This plant sometimes produced simple spores.

HABITAT *Hexagonocaulon* lived in damp, humid conditions.

Typical height 3 in (8 cm)

carbonaceous film

simple branching plants

Hexagonocaulon minutum Lacey & Lewis; Williams Point beds; Triassic; Antarctica.

Range: Triassic–Cretaceous	Distribution: Southern hemisphere	Occurrence: ⌖

SPHENOPSIDS

COMMONLY KNOWN AS HORSETAILS, these plants have jointed stems and leaves produced in whorls about the stem. The spores are produced in spore capsules, which are tightly grouped into cones.

Group: EQUISETALES	Subgroup: EQUISETACEAE	Informal name: Horsetail

Equisetites

These small horsetails had ribbed stems, with spores present in terminal cones. Individual leaves were jointed, linear, or grasslike, with up to 30 per whorl. Scale leaves clothed the base of the leaf. The plant spread by underground stems and tubers.

HABITAT *Equisetites* lived in wetland conditions. Their tubers are often found in fossil soils.

Typical height 20 in (50 cm)

tubers

root

Equisetites sp.; Wealden beds; Early Cretaceous; UK.

Range: Late Carboniferous–Late Cretaceous	Distribution: Worldwide	Occurrence: ⌖ ⌖

Group: EQUISETALES	Subgroup: CALAMOSTACHYACEAE	Informal name: Horsetail

Calamites

This horsetail was actually tree-sized. The stem was ribbed and derived its strength from an outer cylinder of woodlike tissue (as seen in bamboo), protecting a spongy interior. The branches were arranged in whorls around the thick main stem. The leaves were sword-shaped, with a single, central vein, and arranged in whorls around the branch, which gave a dense, almost ornamental-conifer look to the mature plant.

HABITAT *Calamites* was one of several primitive plant groups to reach gigantic sizes in the Late Carboniferous swamps.
REMARK The name *Asterophyllites* is used for one of the foliage types of *Calamites*.

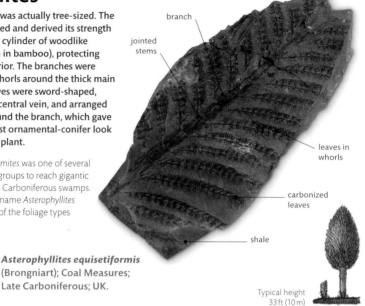

branch
jointed stems
leaves in whorls
carbonized leaves
shale

Asterophyllites equisetiformis (Brongniart); Coal Measures; Late Carboniferous; UK.

Typical height 33 ft (10 m)

Range: Late Carboniferous–Early Permian	Distribution: Worldwide	Occurrence: ✿ ✿ ✿

Group: EQUISETALES	Subgroup: EQUISETACEAE	Informal name: Horsetail

Equicalastrobus

Equicalastrobus is the name given to the isolated "cones" (strobili) of the extinct fossil horsetail *Equisetites*. They are ovoid, circular in cross-section, with helically arranged spore capsules.

HABITAT Modern horsetails thrive in damp, nutrient-poor soils but can survive extreme climates.
REMARK Like the *Aracaucaria* cones (see p.304), these strobili were buried in volcanic ash and many have undergone silicification. This has preserved their internal structures in exquisite detail. Currently, the geological age of these strobili is unknown.

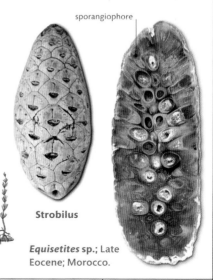

sporangiophore

Strobilus

Equisetites sp.; Late Eocene; Morocco.

Typical height 5 ft (1.5 m)

Range: Late Carboniferous–Late Cretaceous	Distribution: Worldwide	Occurrence: ✿ ✿ ✿

FERNS

THIS LARGE GROUP of plants has a fossil record extending back to the Middle Devonian. Ferns have leaves, called fronds, which usually consist of leaflets, although some have entire or undivided fronds. Reproduction is by spores produced in a variety of spore capsule types, found either on the underside of fronds or, more rarely, produced on specialized fronds. Sizes vary from large tree ferns to minute filmy ferns. They live in a wide range of habitats, ranging from the tropics to cold, temperate regions.

Group: FILICOPSIDA	Subgroup: MARATTIACEAE	Informal name: Fern

Zeilleria

Zeilleria had fronds with minute and extremely narrow leaves. The main leaf stalk subdivided at least four times before the outer frond edge was reached. Spore capsules were found on the lower surface of the frond.

HABITAT Plants of this genus favored swampland.

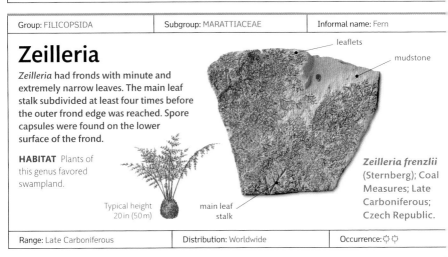

leaflets

mudstone

Typical height 20 in (50 m)

main leaf stalk

Zeilleria frenzlii (Sternberg); Coal Measures; Late Carboniferous; Czech Republic.

Range: Late Carboniferous	Distribution: Worldwide	Occurrence: ✿ ✿

Group: FILICOPSIDA	Subgroup: MARATTIACEAE	Informal name: Tree fern

Pecopteris

Most of these extremely large fronds broke up before fossilization and only toothed fragments, such as those shown here, are found. The leaflets were subdivided two or three times and were typically parallel sided with rounded ends. Venation was usually simple, consisting of a central vein with secondary veins coming at right angles.

HABITAT In life, these plants were tree ferns growing in elevated areas of coal-forming swampland.
REMARK In this specimen, the main stalk is absent.

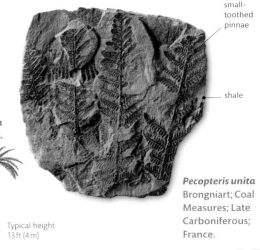

small-toothed pinnae

shale

Typical height 13 ft (4 m)

Pecopteris unita Brongniart; Coal Measures; Late Carboniferous; France.

Range: Late Carboniferous–Early Permian	Distribution: Worldwide	Occurrence: ✿ ✿ ✿

| Group: OSMUNDALES | Subgroup: OSMUNDACEAE | Informal name: Royal fern |

Osmunda

This is a fern with a short stem, large ordinary fronds, and specialized spore-bearing fronds. The stem, where present, has a fibrous structure with many leaf bases passing through it to the outside. The ordinary fronds bear large leaflets with complex netted venation. Spore-bearing fronds either have no leaflets or they are only present at the leaf-stalk end.

Osmunda dowkeri (Carruthers); Thanet Formation; Paleocene; UK.

HABITAT It is probable that, like this modern counterpart, fossil plants of this genus were to be found near water, often in tropical or warm, temperate, wetland areas.

silicified preservation

leaf bases

Typical height 6½ ft (2 m)

Section of stem

| Range: Cretaceous–Recent | Distribution: Worldwide | Occurrence: ✿ |

| Group: FILICALES | Subgroup: DICKSONIACEAE | Informal name: Fern |

Onychiopsis

This plant is known only from its fragmentary fronds. These appear as very delicate-looking, feathery toothed leaves. The leaflets are extremely narrow and join the stem at an angle of about 30 degrees. Spore capsules have been found on fronds from Britain and the US.

Onychiopsis psilotoides (Stokes & Webb); Wealden Beds; Early Cretaceous; UK.

HABITAT *Onychiopsis* grew near lakes, probably in forest-edge habitats, among other ferns, cycads, and cycadlike plants. All ferns require a moderately damp habitat for reproduction.
REMARK At one time, this plant was regarded as a club moss or lycopod, but it is currently thought to be a fern. This specimen is preserved in siltstone.

siltstone

delicate frond structure

Typical height 20 in (50 m)

Frond

| Range: Cretaceous | Distribution: Northern hemisphere | Occurrence: ✿ ✿ |

LYCOPODS

THE LYCOPODS OR CLUB MOSSES have a long fossil history, stretching back to the Late Silurian. They reached their peak in the Late Carboniferous, and today they are represented by only a handful of genera. All plants have helically arranged leaves with spore capsules in the leaf axil, or aggregated into distinct terminal cones. In the Carboniferous, many lycopods were tree-sized, with branches clothed in long, grasslike foliage and cones containing spores. Today, lycopods are small herbaceous plants.

Group: LEPIDODENDRALES	Subgroup: LEPIDODENDRACEAE	Informal name: Giant club moss

Lepidodendron

These tree-sized lycopods or club mosses are notable for their scalelike bark. From the base upward, the plant was anchored in shallow soil by several "Y"-branch rooting organs called *Stigmaria*. These had spirally arranged, finger-sized roots coming from them. A polelike trunk, unbranched for most of its length and up to 130 ft (40 m) or more in height, supported a crown of simple branches. Much of the trunk was covered in diamond-shaped leaf bosses, the familiar *Lepidodendron* fossil. Grasslike leaves, spirally arranged, clothed the upper branches, terminating in cigar-shaped cones (*Lepidostrobus*). These contained, depending on the species, only small spores, large spores, or both.

HABITAT *Lepidodendron* grew in hot and humid swampland.

scalelike surface

Bark

Lepidodendron aculeatum Sternberg; Coal Measures; Late Carboniferous; UK.

ironstone nodule

diamond-shaped pattern

cast

Typical height 100 ft (30 m)

Lepidostrobus variabilis Lindley & Hutton; Coal Measures; Late Carboniferous; Locality Unknown.

Cigar-shaped cone

Range: Carboniferous	Distribution: Worldwide	Occurrence: ✿ ✿ ✿ ✿

Group: DREPANOPHYCALES	Subgroup: DREPANOPHYCACEAE	Informal name: Club moss

Baragwanathia

These were prostrate or low-growing, soft-bodied (herbaceous) plants, with a simple "Y"-branch structure. The stems were always entirely clothed in fine leaves about ⅜ in (1 cm) in length. Spore capsules were present where the leaf joined the main stem (axil). The capsules were organized into zones up the stem, but not into cones.

HABITAT *Baragwanathia* grew in lowland areas and floodplains.
REMARK Much debate has arisen from *Baragwanathia* appearing in the fossil record as early as the Late Silurian, given that it is such an "advanced" plant in comparison to its contemporaries.

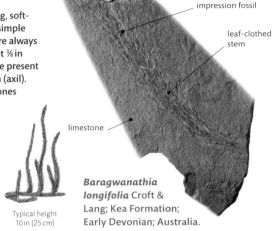

impression fossil

leaf-clothed stem

limestone

Typical height 10 in (25 cm)

Baragwanathia longifolia Croft & Lang; Kea Formation; Early Devonian; Australia.

Range: Late Silurian–Early Devonian	Distribution: Southern hemisphere	Occurrence: ☿

PTERIDOSPERMS

THIS DIVERSE GROUP of plants was at its height during the Late Paleozoic and most of the Mesozoic. Pteridosperms were popularly called seed ferns, after the Carboniferous forms that had foliage seemingly indistinguishable from some ferns. Evidence of seed association, discovered earlier this century, has now placed them within their own group.

Group: MEDULLOSALES	Subgroup: MEDULLOSACEAE	Informal name: Seed fern

Paripteris

The illustrated pollen-bearing organ of *Paripteris* is called a potoniea. Bell-shaped, it was produced upon separate candelabralike structures at the base of the frond. Each potoniea had many fingerlike pollen-producing structures.

HABITAT *Paripteris* was an inhabitant of elevated regions of hot, humid swampland.

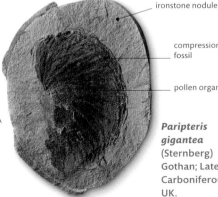

ironstone nodule

compression fossil

pollen organ

Typical height 16½ ft (5 m)

Paripteris gigantea (Sternberg) Gothan; Late Carboniferous; UK.

Range: Late Carboniferous	Distribution: Worldwide	Occurrence: ☿

| Group: MEDULLOSALES | Subgroup: MEDULLOSACEAE | Informal name: Seed fern |

Medullosa

These were seed-bearing plants, growing up to 16½ft (5 m) in height. The stem or trunk consisted partially of old leaf bases, similar to the sago palm (cycad), with prop roots coming off near the base. They bore enormous fronds of various types: some toothed, others with rounded leaflets. All reproduced by seeds, which were often quite large.

HABITAT This was a typical plant of Late Carboniferous swamps.
REMARK The genus *Medullosa* may, in fact, represent a number of similar genera.

Medullosa noei Steidtmann; Coal Measures; Late Carboniferous; US.

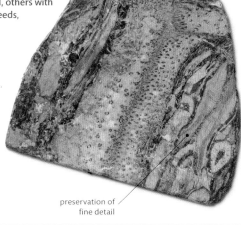

Coal ball section

Typical height 16½ft (5 m)

preservation of fine detail

| Range: Late Carboniferous–Early Permian | Distribution: Worldwide | Occurrence: ✿ ✿ ✿ |

| Group: MEDULLOSALES | Subgroup: MEDULLOSACEAE | Informal name: Seed fern |

Alethopteris

This plant was characterized by large fronds with toothed leaflets. Typically, the thick, robust individual leaflets were not separated from one another, but connected by leaf tissue running between them. Venation was simple: a central vein with smaller veins coming off it at, or near to, right angles.

HABITAT *Alethopteris* grew in elevated areas of hot swamps.
REMARK This genus name applies to the foliage, which was borne on *Medullosa* stems. The name *Alethopteris* is used for a number of plants with a similar frond shape, which may not, in fact, be closely related.

Alethopteris serlii Brongniart; Coal Measures; Late Carboniferous; US.

cast

thick, strong-veined leaflets

ironstone nodule

Typical height 16½ft (5 m)

frond part

| Range: Late Carboniferous–Early Permian | Distribution: Worldwide | Occurrence: ✿ ✿ ✿ ✿ |

| Group: GLOSSOPTERIDALES | Subgroup: GLOSSOPTERIDACEAE | Informal name: Gondwana tree |

Glossopteris

These were tree-sized pteridosperms with rosettes of small to very large leaves. The leaves varied in shape from very narrow to broad and were similar in shape to banana leaves. The venation consisted of a broad central vein which was made up of many smaller veins. From this, a fine reticulum (or net) of minor veins spread, dividing the leaf surface into small lozenge shapes. The wood was of softwood type, with evidence of growth rings, indicating a seasonal climate. Fructifications were borne on specialized leaves, such as pollen-bearing capsules or seed-bearing structures.

HABITAT *Glossopteris* grew in warm, damp lowlands.

Slab of
leaf bed

Glossopteris
sp.; Red beds;
Permian; India.

shale

net venation

sword-shaped
leaves

straight leaf
margins

Typical height
26 ft (8 m)

| Range: Permian | Distribution: Southern hemisphere | Occurrence: ☼ ☼ |

| Group: MEDULLOSALES | Subgroup: MEDULLOSACEAE | Informal name: Seed fern |

Mariopteris

The large fronds of *Mariopteris* had a distinct organization, the pointed leaflets being distributed on a frond divided into four areas. Although not usually found entire, part of one quadrant may be found with a typically reduced basal set of leaflets. Individually, the leaflets had a single central vein, from which secondary veins were given off at an angle of about 60 degrees. The stem consisted in part of old leaf bases, with strongly marked striations along its length.

HABITAT An inhabitant of Late Carboniferous swamps, some species of *Mariopteris* may have been scrambling or climbing plants.

small, carbonized leaflets

Mariopteris maricata (Schlotheim); Coal Measures; Late Carboniferous; Czech Republic.

Typical height 16½ ft (5 m)

Compression fossil

| Range: Late Carboniferous–Early Permian | Distribution: Worldwide | Occurrence: ✿ ✿ ✿ |

| Group: MEDULLOSALES | Subgroup: MEDULLOSACEAE | Informal name: Seed fern |

Trigonocarpus

Casts and compressions of seeds from a variety of pteridospermous plants are found as fossils. This seed had three ribs dividing it into equal sections (and internally into valves). The outside of the seed was usually covered in longitudinal striations representing fibrous or woody bands, which strengthened the seed wall. The seeds were attached terminally or beneath a frond, but are rarely found in their original place in fossil form.

HABITAT *Trigonocarpus* inhabited the less wet areas of hot, humid swamps.

"plumb-bob" shape

Isolated seeds

Typical height 16½ ft (5 m)

Trigonocarpus adamsi Lesquereux; Coal Measures; Late Carboniferous; US.

seeds strongly ribbed

| Range: Late Carboniferous | Distribution: Worldwide | Occurrence: ✿ ✿ |

| Group: PELTASPERMALES | Subgroup: CORYSTOSPERMACEAE | Informal name: Seed fern |

Dicroidium

This large plant, a notable constituent of some Triassic floras of the southern hemisphere, was actually of shrub to small-tree stature. The fronds had unusual "Y"-forked branches, with opposite leaflets along the complete length. Paleobotanists know little about the plant as a whole; however, by association with other plant fossils, they believe it was likely to have been seed bearing.

HABITAT *Dicroidium* was an inhabitant of tropical tree-fern forests.

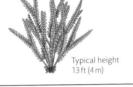

Typical height
13 ft (4 m)

Dicroidium sp.; Ipswich Series; Triassic; Australia.

opposite pairs of leaflets

mudstone

"Y"-forked leaf

| Range: Triassic | Distribution: Southern hemisphere | Occurrence: ✿ ✿ |

| Group: PELTASPERMALES | Subgroup: UMKOMASIACEAE | Informal name: Seed fern |

Pachypteris

A common constituent of many Jurassic floras, *Pachypteris* had small fronds with a regularly or irregularly lobed appearance. It possessed an extraordinarily thick, waxy outer covering, which is often preserved in fossils. It was one of the last of the pteridosperms and became extinct during the Cretaceous period.

HABITAT *Pachypteris* grew in salt marshes.

Pachypteris sp.; Shemshak Formation; Jurassic; Iran.

Typical height
6½ ft (2 m)

shale block

Leaf bed

smallest, lobed leaflets

compression fossil

| Range: Triassic–Cretaceous | Distribution: Worldwide | Occurrence: ✿ ✿ |

BENNETTITES

THE BENNETTITES or cycadeoids were very similar in appearance to sago palms or cycads. They were distinguished by their star-shaped flowers, of which several types are known. The form of the plants varied from globular stumps to branching trees, all with palmlike foliage.

Group: BENNETTITALES	Subgroup: WILLIAMSONIACEAE	Informal name: Cycadeoid

Williamsonia

This plant resembled a shrub or small tree, with diamond-patterned bark and palmlike leaves. Its most interesting aspect was its large, star-shaped flowers. One type contained spore capsules, while the other produced seeds.

HABITAT *Williamsonia* grew in tropical tree fern forests.

Typical height 10 ft (3 m)

carbonized flower

Williamsonia gigas (Lindley & Hutton); Estuarine Series; Middle Jurassic; UK.

Range: Jurassic–Cretaceous	Distribution: Worldwide	Occurrence: ☿

PROGYMNOSPERMS

THIS GROUP IS BELIEVED to be ancestral to all seed plants. Appearing in the Devonian, some were moderately large trees, while others were like tree ferns. They began to produce both small and large spores, and determined spore and seed reproduction during the Late Devonian.

Group: ARCHAEOPTERIDALES	Subgroup: ARCHAEOPTERIDACEAE	Informal name: Archaeopteris

Archaeopteris

Archaeopteris was one of the first trees on Earth. Trees of this genus were small to medium in size, with leafy foliage reminiscent of some conifers. The leafy shoots occurred in opposite arrangement in a single plane. The leaves overlapped one another and were subcircular to nearly wedge shaped. Leaves were replaced by spore capsules on fertile branches.

HABITAT This genus grew in floodplain woodland.

Typical height 33 ft (10 m)

compression fossil

leafy shoots off axis

limestone

Archaeopteris sp.; Kiltorcan Beds; Late Devonian; US.

Range: Devonian	Distribution: Worldwide	Occurrence: ☿ ☿

CORDAITALES

MEMBERS OF THIS GROUP of plants were ancestors of the conifers. They were distinguished by long, leathery leaves, spirally arranged, with many parallel veins. Their reproductive structures were loosely aggregated cones, producing flat seeds with a membranous outer "skirt."

Group: CORDAITANTHALES	Subgroup: CORDAITACEA	Informal name: Cordaite

Cordaites

An ancestor of the true conifers, *Cordaites* was a tree-sized plant that bore huge, strap-shaped leaves in tight spirals around the stem. The leaves have distinctive linear venation. The cones were loose, terminal aggregations of spore capsules or seeds. Some members of the genus were thought to be mangroves, with arching stilt roots.

HABITAT *Cordaites* was an inhabitant of mangrove swamps and elevated hummocks.

straplike leaves

Cordaites angulostriatus
Grand Eury; Coal Measures; Late Carboniferous; UK.

Typical height 33 ft (10 m)

Range: Late Carboniferous–Early Permian	Distribution: Worldwide	Occurrence: ✿✿

CONIFERS

THESE ARE USUALLY shrubs or trees distinguished by the production of woody cones of various shapes and sizes. The leaves are often needle-shaped in temperate and subtropical genera but are flat and broad in tropical podocarps. The seeds are produced mainly in cones.

Group: CONIFERALES	Subgroup: PINACEAE	Informal name: Pine

Pityostrobus

The form-genus represents the cones of various pinelike trees common in the Jurassic and Cretaceous periods. They were about three times longer than broad. The cones were woody, with the curved to reflexed cone leaves or bracts arranged helically about a central axis. Seeds were attached to the bract bases and fell out of the cone when mature.

Pityostrobus dunkeri (Mantell) Seward; Wealden Beds; Early Cretaceous; UK.

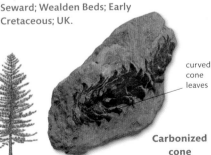

curved cone leaves

Carbonized cone

HABITAT *Pityostrobus* grew in subtropical forests.

Typical height 65 ft (20 m)

Range: Jurassic–Cretaceous	Distribution: Worldwide	Occurrence: ✿✿

Group: CONIFERALES	Subgroup: TAXODIACEAE	Informal name: Coast redwood

Sequoia

This genus includes very large trees also known as coast redwoods. Small, very globular cones are a feature of these plants. The woody cone leaves (or bracts) are arranged helically about a central axis and have a seed with a scale-leaf on the upper surface, near the base or point of attachment. The cones do not often disintegrate (as disassociated bracts), even as fossils, but open to let the seeds fall out.

HABITAT Redwoods once formed extensive forests in subtropical regions of the world. These forests, as found today in California, are of an open nature, with few other tree species present.

globular
shape

Pine cones

ironstone
carbonized
interior

Sequoia dakotensis
Brown; Hell Creek Formation; Late Cretaceous; US.

Typical height 230 ft (70 m)

Range: Jurassic–Recent	Distribution: Worldwide	Occurrence:

Group: CONIFERALES	Subgroup: TAXODIACEAE	Informal name: Wellingtonia

Sequoiadendron

These giant redwood trees are not as tall as *Sequoia*, but are distinguished by the huge diameter of the trunk. They have small leaves or needles, in common with many other conifers. The leaves are scalelike, arranged alternately along either side of the twig. They are very small, with a blunt, toothlike shape, and are pressed closely to the stem. Wood from the tree is of softwood or gymnospermic type, with no conducting tubes but often with growth rings.

HABITAT Trees of this genus grew in the same subtropical forest habitats as the *Sequoia*.
REMARK Counting the rings shows that the plants can live to impressive ages. Over 1,000 years is by no means uncommon for the larger examples seen growing in California.

Typical height 200 ft (60 m)

compression fossil
leaves in
opposite
pairs
small
leaves
shale

Sequoiadendron affinis
(Lesquereux); Oligocene; US.

Range: Jurassic–Recent	Distribution: Worldwide	Occurrence:

| Group: CONIFERALES | Subgroup: Unclassified | Informal name: Softwood |

Conifer wood

Silicified conifer wood; Lower Greensand; Early Cretaceous; UK.

Conifer wood is distinguished by its very uniform appearance and is commonly called "softwood." Growth rings can often be seen, as can resin canals or ducts. The bulk of the tissue is made up of a single type of woody cell called a tracheid. Generic and species distinctions between softwoods is difficult and can usually only be made at microscopic level from the study of ground sections. Similar wood is produced by *Ginkgo* and by some progymnosperms and pteridosperms.

HABITAT Conifer wood is found in a wide selection of plant community types, including tropical to temperate forests, woodlands, parklands, and mangroves.

dark-stained outer layer

Typical height 100 ft (30 m)

wide growth rings

silicified preservation

| Range: Permian–Recent | Distribution: Worldwide | Occurrence: ✿ ✿ ✿ ✿ |

| Group: CONIFERALES | Subgroup: TAXODIACEAE | Informal name: Dawn redwood |

Metasequoia

Metasequoia occidentalis (Newberry) Chaney; Shale; Oligocene; Canada.

These are large trees with small leaves in the form of flattened needles, in opposite pairs along the twig. Seed cones are produced upon long stalks and the cone leaves or bracts are in cross-pairs. At the base of the tree, where it joins the roots, vertical growths of wood called "knees" are produced.

HABITAT *Metasequoia* grows in subtropical swampland, thriving in very wet areas with its roots immersed in water.
REMARK The plant was known only as a fossil until 1941, when it was found growing in Szechuan, China—a remarkable example of a "living fossil."

short, featherlike shoots

carbonized preservation

Typical height 100 ft (30 m)

mudstone

leaves in opposite pairs

| Range: Cretaceous–Recent | Distribution: Northern hemisphere | Occurrence: ✿ ✿ |

Group: CONIFERALES	Subgroup: ARAUCARIACEAE	Informal name: Monkey puzzle

Araucaria

This genus of large trees includes the Monkey Puzzle Tree and Norfolk Island Pine. The leaves are small and toothlike, spirally arranged, and closely pressed to the twig, hiding it completely. The cones are large, almost spherical, and very spiny. The surface is formed of the closely packed woody ends of cone leaves or bracts. Seeds are present at the base of the bracts.

HABITAT This genus grew in subtropical mountain forests. Recent genera are restricted to the southern hemisphere.

REMARK The famous fossil forest buried by volcanic ash at Cerro Cuadrado, Patagonia, is made up of *Araucaria* trees. Silicified twig, cone, and wood remains from there are common.

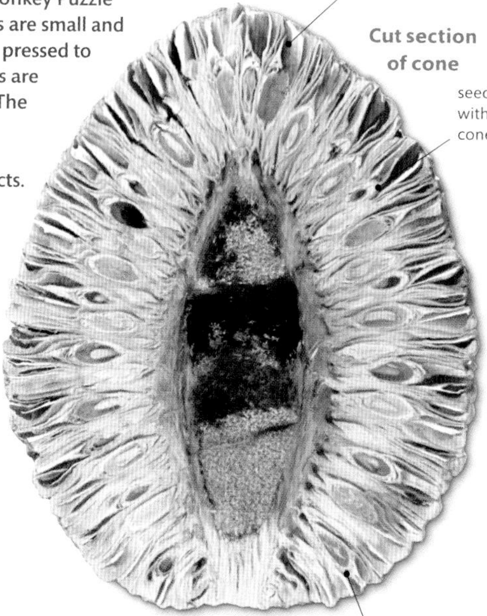

all tissues preserved

Cut section of cone

seeds within cone

woody surface

silicified preservation

Side view

broken surface

interlocking diamond pattern

Top view

rusty color

seeds in diamond-shaped cavities

Araucaria mirabilis (Spegazzini) Windhausen; Cerro Alto Beds; Early Cretaceous; Argentina.

Typical height 100 ft (30 m)

Range: Jurassic–Recent	Distribution: Worldwide	Occurrence: ✿ ✿

Group: CONIFERALES	Subgroup: UNCLASSIFIED	Informal name: Softwood

Silicified wood

Petrification of wood involves the replacement of the original tissue by minerals in solution, usually silica or calcium carbonate. Recrystallization of silica due to pressure and heat, together with impurities in the solution, causes new forms of this mineral—usually agate or, more rarely, opal—to grow. It also results in the destruction of the wood anatomy.

HABITAT Silicified wood is found in a very wide variety of habitats.
REMARK It is often impossible to attribute silicified or similarly preserved woods to particular plants or trees. All that can be said is that the wood is of a broadly soft wood type (as illustrated). Silicified wood is often attractive and multicolored.

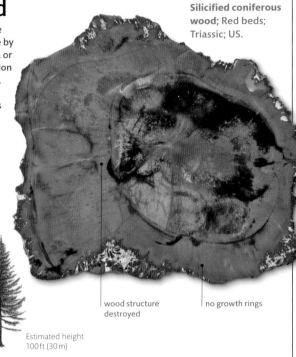

Silicified coniferous wood; Red beds; Triassic; US.

wood structure destroyed

no growth rings

Estimated height 100 ft (30 m)

Range: Permian–Recent	Distribution: Worldwide	Occurrence: ✿ ✿

Group: CONIFERALES	Subgroup: PINACEAE	Informal name: Spruce

Picea

This is a large group of mainly tall, narrowly conic trees that today is confined to the northern hemisphere. Many spruce trees have characteristic drooping branches, giving a distinctive tiered effect to the whole plant. The stiff, needlelike leaves are arranged spirally about the stem and are ¾–1¼ in (2–3 cm) long. They grow from small pegs, which remain if the leaves are shed. The woody cones are pendulous, ovoid, or cylindrical and often have strongly recurved bracts, which open to shed seeds as the cone matures.

HABITAT The wood is of softwood type, often with strongly marked growth rings, in keeping with the seasonal montane climate found in most pine habitats.

Carbonized cone

Picea sp.; Thanet Formation; Late Paleocene; UK.

woody bracts

diamond pattern

Typical height 130 ft (40 m)

Range: Early Cretaceous–Recent	Distribution: Worldwide	Occurrence: ✿ ✿

Group: CONIFERALES	Subgroup: Not applicable	Informal name: Amber

Amber and Copal

Amber is fossilized resin or gum produced by some fossil plants. The earliest recorded fossil resins are of Carboniferous age, but ambers do not occur until the Early Cretaceous. Famous amber deposits include those from the Baltic region and the Dominican Republic. It is likely that ambers were mainly ancient gymnosperm (probably conifer) resins, but today such gums are also produced by flowering plants. Baltic amber occasionally contains insect and plant remains. It is supposed to have been formed in forests of a primitive species of pine, *Pinus succinifera*. Recent and semifossil copal resins differ from amber in that they are still readily soluble in organic solvents. Resins are exuded from within the tree when it is traumatized due to attack or growth splits. Today they are gathered commercially, one example being the copals derived from Kauri pine in New Zealand. Baltic amber is used in jewelry, and copals in varnish manufacture.

Kauri gum; Pleistocene; New Zealand.

stalactitic flow

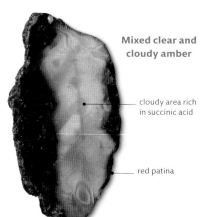

Mixed clear and cloudy amber

cloudy area rich in succinic acid

red patina

Baltic amber (Succinite); Late Eocene; Denmark.

Clear gem-quality amber

Carved Dominican amber

Tree (*Pinus* sp.)

Typical height 100 ft (30 m)

Range: Early Cretaceous–Recent	Distribution: Worldwide	Occurrence: ✿

GINKGOS

THIS WAS FORMERLY an extensive group of plants, but it is now represented by a solitary relict genus, the *Ginkgo* or maidenhair tree. The group appeared during the Early Permian and was at its height during the Jurassic. All members had characteristic fan-shaped leaf architecture.

Group: GINKGOALES	Subgroup: GINKGOACEAE	Informal name: Maidenhair tree

Ginkgo

These very tall trees have a distinctive fan- or wedge-shaped foliage. The leaves may be notched or almost entire. The trees produce spherical seeds.

HABITAT A "living" fossil, the *Ginkgo*'s natural habitat is China, but it has been introduced as an ornamental tree in many parks and gardens.

Typical height 115 ft (35 m)

Ginkgo sp.; Ardtun Leaf Bed; Paleocene; UK.

fan-shaped leaf

compression fossil

mudstone

Range: Late Triassic–Recent	Distribution: Worldwide	Occurrence: ✿ ✿

EUDICOT ANGIOSPERMS

EUDICOTS ARE A MAJOR GROUP of angiosperm flowering plants that have two seed leaves and a particular pollen structure. The leaves range from herbaceous to tree types. All reproduce sexually by seeds. The earliest record of the group is from Early Cretaceous pollen.

Group: PLATANALES	Subgroup: UNKNOWN	Informal name: Extinct sycamore

Betulites

The leaves of this genus have some similarity to those of *Betula* or the birch family. They were round, oval, or heart-shaped and always had teeth along the margin. Venation consisted of a central vein, with secondary veins coming off at about 45 degrees. They are usually found disassociated from twigs, which indicates that at least some were dropped seasonally.

HABITAT This genus grew in temperate climates and usually damp habitats.

Typical height 33 ft (10 m)

ironstone nodule

heart-shaped leaf

Betulites sp.; Dakota Sandstone; Late Cretaceous; US.

Range: Late Cretaceous–Miocene	Distribution: Worldwide	Occurrence: ✿ ✿

Group: PLATANALES	Subgroup: Unclassified	Informal name: Extinct plane

Araliopsoides

This genus is characterized by large leaves. Typically, they were coarsely three-lobed but were sometimes of very variable appearance. The leaf margins were smooth with no teeth or serrations. The venation was simple—a central vein with straight secondary veins coming off at about 45 degrees. They were very thick and resilient, which explains their survival in sandstones.

HABITAT *Araliopsoides* grew in warm, temperate to subtropical, deciduous forests.

simple venation

three-lobed leaf

Sandy-ironstone cast

Typical height 33 ft (10 m)

Araliopsoides cretacea (Newberry); Dakota Sandstone; Late Cretaceous; US.

Range: Late Cretaceous	Distribution: Worldwide	Occurrence: ✿

Group: SAPINDALES	Subgroup: ACERACEAE	Informal name: Maple

Acer

These are small to large trees of temperate regions, popularly called maples. The leaves are usually of a basic three-lobed appearance, toothed, with a long stalk, and are shed in the fall. The wood is zoned, with noticeable springwood composed of large conducting tubes. Flowers are produced in drooping clusters and are followed by the distinctive winged fruits. These are always produced in pairs, but they can become separated when they fall to the ground.

HABITAT This genus is an inhabitant of temperate, deciduous forests.

***Acer* sp.;** Freshwater limestone; Miocene; Croatia.

single wing with venation

limestone

Winged fruit

Typical height 80 ft (25 m)

Range: Oligocene–Recent	Distribution: Worldwide	Occurrence: ✿

Group: UNKNOWN	Subgroup: UNKNOWN	Informal name: Fig

Ficus

A large genus of shrubs and trees, the figs have thick, oval to fiddle-shaped leaves, with a noticeable central vein. The fruit is a globular, flat-topped, and stalked structure, with seeds embedded in a fleshy surround.

HABITAT This is a tropical to temperate genus.
REMARK The single quotation marks around the name *Ficus* indicate that it is the closest genus available but that there is some doubt that it has been correctly identified.

'Ficus' sp.; Early Eocene; US.

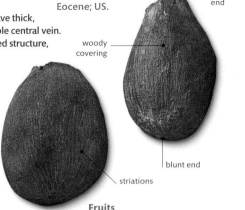

pointed end

woody covering

blunt end

striations

Fruits

Typical height 100 ft (30 m)

Range: Eocene–Recent	Distribution: Worldwide	Occurrence: ✿

Group: FAGALES	Subgroup: FAGACEAE	Informal name: Oak

Quercus

A genus of large trees with distinctive foliage and fruits, the leaves of *Quercus* are a variety of shapes, from lobed to fiddle-shaped or hollylike. The leaves may be shed in the fall or they may be evergreen. The fruit is the familiar acorn, with the ornament of the cup varying from species to species. Usually the wood has very large conducting tubes and distinct growth rings.

Quercus sp.; Miocene; US.

HABITAT *Quercus* grows in temperate, deciduous to semievergreen forests and woodlands.
REMARK Although *Quercus* has been recorded from the Late Cretaceous, the oldest substantial pollen based records are from the Paleocene of Europe. Typical oak foliage, acorns, and wood are known from the Middle to Late Eocene of North America.

distinct ray structure

silicified wood

growth rings

mineral impurities cause coloration

Typical height 130 ft (40 m)

Range: Paleocene–Recent	Distribution: Worldwide	Occurrence: ✿ ✿

| Group: SAXIFRAGALES | Subgroup: HAMAMELIDACEAE | Informal name: Sweetgum tree |

Liquidambar

These small to large trees are often called sweetgum. The twigs sometimes have corky flanges or wings, and the leaves are distinctively lobed or star-shaped, with five to seven points and toothed margins. Leaf venation consists of a single, central vein for each lobe, with nets of veins on either side. Leaves are shed in the fall. Fruits are aggregates of woody, burrlike capsules, borne on a long stem. The wood is fine grained, with small conducting tubes.

HABITAT They grow in temperate, deciduous to semievergreen forests and woodlands.

Liquidambar europeanum Braun; Freshwater limestones; Miocene; Switzerland.

limestone

five-lobed leaf

simple venation

carbonaceous film

Typical height 80 ft (25 m)

| Range: Oligocene–Recent | Distribution: Worldwide | Occurrence: ☼ |

| Group: FAGALES | Subgroup: BETULACEAE | Informal name: Birch |

Betula

This is a small tree, with bark that is often distinctively patterned or colored in modern representatives. The leaves are usually small with toothed margins, and they are shed during the fall. Trees of this genus reproduce with catkins and seeds, with seed leaves or bracts. The wood is fine grained, with very small conducting tubes and sometimes with complex wood rays.

HABITAT *Betula* was a common component of northern ice-age forests. Today it is found in temperate climates, often in waterside habitats.

Betula sp.; Brown coal; Miocene; Germany.

mudstone

toothed leaf margins

simple venation

Typical height 50 ft (15 m)

| Range: Miocene–Recent | Distribution: Worldwide | Occurrence: ☼ |

MONOCOTYLEDONOUS ANGIOSPERMS

THIS IMPORTANT GROUP of plants is characterized by their mode of germination—with a single seed leaf—and by leaves with parallel venation. Included in the group are grasses, narcissi, sedges, rushes, orchids, lilies, irises, and palms—all of them with a bewildering range of flowers and reproduction strategies. Fossil forms are recognized from the Late Cretaceous as palms and rushes, while grasses do not appear until the Early Cenozoic Era. A possible monocot is recorded from the Triassic of America.

Group: ARECALES	Subgroup: ARECACEAE	Informal name: Palm tree

Palmoxylon

Fossil palm-tree wood is quite homogeneous and typically has many round, dark spots on a lighter background in cross-section. These spots are conducting or vascular tissues and distinguish the genus from dicotyledonous angiosperms and conifer woods. There are no growth rings.

HABITAT *Palmoxylon* grew in subtropical, semievergreen forest and woodland.

Palmoxylon sp.; Mio-Pliocene; Antigua.

visible conducting tissue

Typical height 65 ft (20 m)

Silicified stem

Range: Paleocene–Pliocene	Distribution: Worldwide	Occurrence: ✿ ✿

Group: ARECALES	Subgroup: ARECACEAE	Informal name: Stemless palm

Nypa

This is a palm with little or no stem. The leaves or fronds are large, with long, pointed, subopposite leaflets. Coconutlike fruits are produced near the base of the frond, and within the fibrous husk are several seeds.

HABITAT Stemless palms grow in mangrove swamps and are restricted to southeast Asia.

REMARK These plants were a distinctive feature of some Early and Middle Eocene plant communities.

Nypa burtinii (Brongniart); Sables de Bruxelles; Middle Eocene; Belgium.

subspherical shape

woody outer layer

Typical height 5 ft (1.5 m)

Fruit

Range: Late Cretaceous–Recent	Distribution: Northern hemisphere	Occurrence: ✿ ✿ ✿

GLOSSARY

THE USE OF SOME technical expressions is unavoidable in a book of this nature. Commonly used technical words are often defined in the text itself, and many terms are explained by the annotated illustrations. The definitions given below have been simplified and generalized and are appropriate for this book only. Words in bold type are explained elsewhere in the glossary.

■ **Adductor muscles**
The muscles controlling the valves of bivalves and brachiopods.

■ **Adipose fin**
In fish, a fleshy **dorsal** fin.

■ **Alate plate**
A flaplike extension on the interior hinge line of the **brachial valve** of a brachiopod.

■ **Alveolus**
A tooth socket in vertebrates; a cavity containing **phragmocone** in belemnites.

■ **Ambulacra**
The plated zones of echinoderms associated with tube feet.

■ **Anal fin**
An unpaired **posterior** fish fin.

■ **Anal tube**
The projecting part of the arm that carries the anus in crinoids.

■ **Antennules**
The sensory appendages toward the front of a trilobite.

■ **Anterior**
Toward the front.

■ **Aperture**
The opening surrounded by the shell margin in mollusks.

■ **Apical disk**
The upper central disk of a sea urchin, from which the **ambulacra** radiate.

■ **Aragonite**
A crystalline form of calcium carbonate.

■ **Aristotle's Lantern**
The five-sided feeding and locomotor structure located around the mouth of sea urchins.

■ **Articular bones**
The bones incorporated in the articulation between the skull and lower jaw in some vertebrates.

■ **Auricles**
The earlike structure on a bivalve shell.

■ **Avicularia**
The defensive individuals in a bryozoan colony.

■ **Axial boss**
The raised node at the center of the **corallites** in corals.

■ **Axis**
In trilobites, the ridge that runs from **anterior** to **posterior**, down the midline; in vertebrates, the first or second element of the backbone.

■ **Barbel**
A slim, sensitive process near the mouth of some fish.

■ **Basal**
At or pertaining to the base.

■ **Benthonic**
Living on the sea floor.

■ **Bicipital surface**
A bulbous surface at the **proximal** end of the upper arm bone in vertebrates.

■ **Biconic**
Pointed at both ends.

■ **Bifid**
Divided into two.

■ **Bifurcating**
Dividing into two parts.

■ **Boreal**
Suggesting a fauna with a preference for a cold climate.

■ **Brachial valve**
One of the two valves that form the brachiopod shell.

■ **Brachiole**
An armlike appendage to a brachiopod.

■ **Bract**
A modified leaf associated with plant reproductive structures.

■ **Bunodont**
A condition in which molar teeth possess rounded cusps and tubercles.

■ **Byssal notch**
The notch near the **hinge** of some bivalve shells, associated with attachment threads.

■ **Byssus**
The attachment threads of a bivalve.

■ **Calcareous**
Made of calcium carbonate; chalky.

■ **Calice**
The upper portion of a coral skeleton.

■ **Callum**
The dome of calcite in mollusks.

■ **Calyx**
In crinoids, a cup-shaped structure to which the arms are attached; in plants, the outer circle of a flower made up of sepals.

■ **Capitilum**
A portion of a crustacean or chelicerate, usually composed of armored plates, protecting the food-gathering organs.

■ **Carapace**
In crustaceans and chelicerates, the external shield covering head and trunk; in vertebrates, the upper shell.

■ **Cardinal teeth**
The articulating structures on the hinge of a brachiopod.

■ **Carina (***pl.* **Carinae)**
A ridge- or keel-shaped structure.

■ **Carnassial tooth**
A molar or premolar specialized for cutting.

■ **Caudal fin**
The tail fin of a vertebrate.

■ **Cenozoic**
The fourth era of time in the history of the Earth, 65 to 2 million years ago.

■ **Centrum (***pl.* **Centra)**
The main body of a **vertebra**.

■ **Cephalon**
The head of a trilobite.

■ **Cephalothorax**
The combined head and trunk region of an arthropod.

■ **Cerci**
The sensory appendages at the **posterior** of the abdomen of insects.

■ **Chela (***pl.* **Chelae)**
The pincerlike end of the limb of crustaceans or chelicerates.

■ **Chelicerae**
The biting appendages of spiders.

■ **Chevron bones**
In reptiles, the pair of bones, often fused to form a "Y," that hang below the tail **vertebrae**.

■ **Chitin**
A horny substance, forming all or part of the skeleton of arthropods.

■ **Cilia**
Tiny, hairlike structures on a cell.

■ **Cirri**
A tendril-like animal appendage; the jointed **thoracic** appendages of barnacles.

■ **Claspers**
The specialized pelvic fins of sharks, rays, and rabbitfishes.

■ **Coenosteum**
The **calcareous** skeleton of a bryozoan and some corals.

■ **Columella**
The column that surrounds the shell's axis in gastropods.

■ **Columnal**
The stem **ossicles** in crinoids.

■ **Commissure**
The line along which two valves of a shell meet.

■ **Compression fossil**
A flattened fossil.

■ **Condyle**
A convex articular surface in some skeletal joints.

■ **Consolidated**
(of sediment) Hardened and/or compacted.

■ **Corallite**
In corals, an individual **polyp's** skeleton.

■ **Cornua**
A hornlike structure in fish.

■ **Costae**
Fine, concentrically arranged ribbing.

■ **Cotyle**
The concave articular surface in some skeletal joints.

■ **Coxa (pl. Coxae)**
The proximal or base region of a limb, which articulates with the body in insects and some other arthropods.

■ **Cranium**
The part of the skull enclosing the brain.

■ **Cuticle**
The hardened outer surface of the external skeleton of arthropods.

■ **Cyst**
A thick-walled cell in plants.

■ **Degenerate**
Not fully formed.

■ **Deltoid**
A "V"-shaped plate in the cup of a blastoid.

■ **Deltoid crest**
The projection of the upper arm bone for the attachment of muscles in vertebrates.

■ **Demosponge**
A sponge with a skeleton possessing one- to four-rayed sponge fibers.

■ **Dentary**
The bone of the lower jaw in vertebrates.

■ **Denticle**
A small, toothlike structure.

■ **Denticulate**
Serrated; housing **denticles**.

■ **Dermal armor**
The bony plates situated in the skin of some vertebrates.

■ **Dermal denticle**
A toothlike structure found in the skin of sharks.

■ **Detritus**
Fine particles of organic matter.

■ **Diagenesis**
The process involved in turning sediment into a rock.

■ **Diastema**
The space between two types of teeth.

■ **Dissepiment**
In corals, a vertical plate located between **septa**.

■ **Dorsal**
The upper surface or back.

■ **Dorsoventrally flattened**
Compressed from top to bottom.

■ **Elytra**
The modified **anterior** wings that act as protective covers for the membranous hindwings in beetles.

■ **Enameloid**
Mineralized dental tissue similar to mammalian enamel.

■ **Entoplastron**
The **anterior**, median bony plate of the lower shell in turtles.

■ **Epidermis**
The bloodless, nonsensitive portion of the skin in vertebrates.

■ **Epiphyseal plate**
The plate from which elongation (growth) takes place in a mammal's long bone.

■ **Epiplanktonic**
The area of the sea from the surface to about 100 fathoms.

■ **Epiplastron**
One of the anterior pair of bony plates of the lower shell in turtles.

■ **Epistome**
The area covering the mouth and second antennae, and the plate covering this region, in crustaceans.

■ **Escutcheon**
A depression found behind the **umbones** of a bivalve shell.

■ **Evolute**
Loosely coiled.

■ **Exoskeleton**
The hard outer casing of arthropods.

■ **External mold**
An impression of the outside of an organism.

■ **Facial suture**
A line on the head of a trilobite along which splitting occurred during molting.

■ **Fenestrule**
An opening located between the branches of a bryozoan colony.

■ **Filter feeder**
An organism that gains nutrition by filtering particles from the water.

■ **Flagellum**
A whiplike appendage.

■ **Foot**
In invertebrates, the base from which the organism grows or by which it is cemented to the substrate.

■ **Foramen (pl. Foraminae)**
A small opening or perforation in brachiopods.

■ **Form genus**
A genus containing many similar species but which may not actually be related.

■ **Frontal bone**
In vertebrates, one of a pair of bones situated in the skull.

■ **Fusiform**
Tapering at each end.

■ **Genal spine**
A trilobite's cheek spine.

■ **Genalangle**
The angle between the back and lateral margins of a trilobite's head.

■ **Girdle**
In chitons, the outer portion of the **mantle**; in vertebrates, the hooplike group of bones that support the limbs.

■ **Glabella**
In trilobites, the central region of the head; in vertebrates, the prominent front bone which joins the ridges above the eyes.

■ **Glauconitic**
Containing the mineral glauconite, the presence of which indicates that a sediment was deposited in marine conditions.

■ **Gonads**
Sex glands.

■ **Gracile**
Lightly built.

■ **Guard**
A massive, bullet-shaped calcite structure in belemnites.

■ **Hadrosaurian crest**
A bony crest found on the skulls of hadrosaurian dinosaurs.

■ **Halteres**
The pair of structures representing hindwings in flies.

■ **Heterocercal tail**
The tail of a fish with unequal-sized lobes.

■ **Hinge plate**
In mollusks, the portion of valve that supports the **hinge teeth**; in brachiopods, the socket-bearing portion of the dorsal valve.

■ **Hinge teeth**
Articulating structures in mollusk shells.

■ **Holdfast**
A basal structure that attaches a plant to the substrate.

■ **Homocercal tail**
A symmetrical fish tail.

■ **Hyoplastron**
The paired, second lateral, bony plate of the lower shell of a turtle.

■ **Hypocercal tail**
A type of fish tail in which the **notochord** ends in the extended lower lobe.

■ **Hypoplastron**
The paired, third lateral, bony plate of the lower shell of a turtle.

■ **Incus**
The central of the three bones of the inner ear of vertebrates.

■ **Interambulacral**
The area between two radial plates along which the tube feet of echinoids are arranged.

■ **Interarea**
The flat area on a brachiopod shell between the **hinge** line and beak areas.

■ **Internal mold**
An impression of the inside of an organism.

■ **Involute**
Tightly coiled.

■ **Ischium**
One of the paired hindlimb girdle bones of vertebrates.

■ **Keel**
A raised midline structure that runs longitudinally along the medial surface of the shell venter in some ammonites.

■ **Labial**
Pertaining to the lips.

■ **Labial spines**
The spines situated on the aperture of certain gastropods.

■ **Lamina**
A thin sheet or layer of tissue.

■ **Lateral teeth**
In bivalves, articulation structures toward the margin of the hinge.

■ **Ligament pit**
Depression housing elastic structure joining valves of a bivalve.

■ **Lithological unit**
The name and/or type of rock.

■ **Living fossil**
An animal or plant species that has remained almost unchanged for millions of years.

■ **Lobolith**
A large globose float at the base of the stem of some Paleozoic crinoids.

■ **Locally occuring**
Found only at certain specific localities.

■ **Loph**
A crest or ridge.

■ **Lunule**
A crescent-shaped structure or mark in the **test** of some sea urchins.

■ **Macroconch**
The larger form of a shell in a species in which males and females differ.

■ **Malleus**
The outermost bone of the inner ear in vertebrates.

■ **Mantle**
The external body wall lining the shell of some invertebrates.

■ **Mantle cavity**
The space between the body and **mantle** of a bivalve.

■ **Marine indicators**
Characteristics denoting saltwater conditions.

■ **Marl**
A **calcareous** mudstone.

■ **Matrix**
The material on which an organism rests or is embedded.

■ **Maxillary lobes**
A part of the mouth of an arthropod.

■ **Median tooth**
A tooth placed in the midline of the mouth.

■ **Mesial tooth**
A tooth placed near the middle of the jaw.

■ **Mesoplastron**
In turtles, one of a pair of bony plates forming the lower shell.

■ **Mesosuchian**
A primitive crocodile.

■ **Mesozoic**
The second era of the Phanerozoic eon, from 248 to 65 million years ago.

■ **Metamorphism**
A change in an organism's structure during development.

■ **Metatarsal**
Pertaining to the bones in the ankle.

■ **Microconch**
The smaller form of a shell in a species i which males and females differ in size.

■ **Microsculpture**
Small patterns of ornamentation.

■ **Mold**
An impression obtained from an original form.

■ **Montane climate**
Hilly areas below the timberline.

■ **Monticule**
A hummock on the surface of a bryozoan colony.

■ **Morphology**
The structure and form of plants and animals.

■ **Mucillage**
A carbohydrate found in certain plants, which can be secreted.

■ **Mucron**
A short tip or **process**.

■ **Mucronate**
Ended by a short tip or process.

■ **Nacre**
An iridescent internal layer of a mollusk shell.

■ **Nacreous**
Made of nacre.

■ **Nema**
The attachment thread found in some graptolites.

■ **Neotony**
A condition in which development is terminated at a preadult stage but sexual maturity is reached.

■ **Neural arch**
The upper portion of a **vertebra**.

■ **Neural canal**
The channel through which the spinal cord passes.

■ **Neural plate**
A series of bony plates located in the midline of the upper shell of turtles.

■ **Neural spine**
The blade- or pronglike structure located on the dorsal aspect of a **vertebra**.

■ **Nodes**
Bumps or protruberances; in plants, the attachment point of leaf stem.

■ **Notochord**
A skeletal rod located above the nerve chord of fish.

■ **Nutritive groove**
Food channel.

■ **Nymph**
An immature insect or, in bivalves, the narrow ledge on the **hinge** behind the **umbo**.

■ **Occipital**
A bone at the rear of the skull which articulates with the spinal column.

■ **Occlusal**
The cutting or grinding surface of a tooth.

■ **Operculum**
The structure attached to a gastropod's foot, used to close the shell's **aperture**.

■ **Orbit**
An eye socket.

■ **Ornithopod**
A group of dinosaurs with bird-like feet.

■ **Orthocone**
The long, straight shell of a nautiloid cephalopod.

■ **Ossicles**
In invertebrates, the **calcareous** bodies that make up the skeleton.

■ **Ossified**
Turned to bone.

■ **Otolith**
A **calcareous** concretion in a vertebrate's ear.

■ **Paleozoic**
The first era of the Phanerozoic eon, 545 to 248 million years ago.

■ **Pallets**
The burrowing structure of mollusks.

■ **Pallial line**
In bivalves, an impressed line inside the valve, parallel to the margin, caused by the attachment of the mantle.

■ **Pectoral fin**
One of a pair of forward-pointing fins in fish.

■ **Pedicle**
A cuticle-covered appendage for the attachment of a brachiopod shell to the substrate.

■ **Pedipalps**
The first pair of postoral appendages in chelicerates.

■ **Peduncle**
In some invertebrates, the stalk that supports most of the body.

■ **Pelvic fin**
In fish, one of a pair of fins placed toward the back.

■ **Pen**
The horny internal skeleton of squids.

■ **Periderm**
A horny, cuticular covering.

■ **Phragmocone**
In belemnites and other mollusks, the conelike internal shell that is divided into chambers by **septa** and perforated by a **siphuncle**.

■ **Phosphatic**
Composed of phosphate; in fossils, usually calcium phosphate.

■ **Pinnae**
Small leaflets on plants.

■ **Pinnule**
A part of the fanlike structure of the arms of a crinoid.

■ **Pith**
A spongy plant tissue (of plant stem).

■ **Planispiral**
Coiled in one plane.

■ **Plankton**
Weak-swimming or passively floating animals or plants.

■ **Plastron**
The lower bony shell of turtles.

■ **Pleotelson**
In crustaceans and chelicerates, the plate formed by the fusion of tail plates with abdominal segments.

■ **Pleural lobes**
The lateral parts of the **thoracic** segments of trilobites.

■ **Pleural spine**
The body spine of a trilobite.

■ **Pneumatic bone**
In birds, the bones connected by canals to the respiratory system.

■ **Polymorphic**
A species with more than one form.

■ **Polyp**
An individual member of a coral colony.

■ **Posterior**
Toward the rear.

■ **Presacral vertebrae**
Part of the backbone in front of the hindlimb girdle.

■ **Prismatic**
A shell structure of minute, columnar prisms.

■ **Proboscis**
In insects, a tubular feeding structure; in mammals, a flexible, elongated snout.

■ **Process**
An extension or appendage to a organism.

■ **Pronotum**
The back of the first body segment in insects.

■ **Pro-ostracum**
The thin, tonguelike extension in front of the **guard** in belemnites.

■ **Proparian**
A form of suture in trilobites that reaches the edge in front of the genal angle.

■ **Prop root**
A root that helps support a plant.

■ **Proximal**
At or toward the near, inner, or attached end.

■ **Pseudopelagic**
Refers to the lifestyle of organisms that live attached to floating objects in the sea.

■ **Pubic bone**
One of the paired girdle bones in vertebrates.

■ **Pustule**
A small, raised mound or **tubercle**.

■ **Pygidium**
The tail of a trilobite.

■ **Pyrite**
Gold-colored mineral composed of iron sulfide.

■ **Quadrate**
The bone that produces the **articulation** of the skull and lower jaw in some vertebrates.

■ **Radial**
Diverging from the center.

■ **Radula**
A horny or toothlike structure located in the mouth of all mollusks except bivalves.

■ **Ramus**
A **process** projecting from a bone.

■ **Recurved**
Curved backward.

■ **Reflexed**
Turned backward.

■ **Reticulum**
A fine network.

■ **Rhizome**
An underground rootlike stem bearing roots and shoots.

■ **Rock former**
Animals that occur in sufficient abundance and frequency as to form the major bulk of a rock.

■ **Rostral plate**
One of the plates covering a barnacle (also known as a rostrum).

■ **Rugae**
Wrinkles on a shell surface.

■ **Sacculith**
The largest otolith in the inner ear.

■ **Sacral**
Associated with the hip girdle.

■ **Sagittal crest**
The median crest located on the posterior portion of the skull.

■ **Scale leaf**
A tough, membranous leaf that often has a protective function.

■ **Sclerosponge**
A calcified demosponge.

■ **Sclerotic ring**
A ring of bony plates around the eye socket in reptiles and birds.

■ **Scute**
The bony, scalelike structure found in reptiles.

■ **Sedentary**
Not or barely moving.

■ **Selenizone**
A groove in a gastropod shell.

■ **Septum (** *pl.* **Septa)**
A thin dividing wall.

■ **Sexual dimorphism**
A condition in which the sexes differ in form.

■ **Shelf sea**
Shallow seas around land masses.

■ **Sicula**
The cone-shaped skeleton of a graptolite colony.

■ **Siltstone**
A rock formed from silt deposits.

■ **Sinus**
A cavity or recess in gastropods.

■ **Siphon**
A tubular element used for the intake of water in mollusks.

■ **Siphuncle**
A tubular extension of the **mantle** passing through all chambers of shelled cephalopods.

■ **Somite**
A body segment of a crustacean.

■ **Spicule**
A spikelike supporting structure in many invertebrates, especially sponges.

■ **Spinneret**
An organ that spins fiber from the secretion of silk glands.

■ **Spinule**
A small, spinelike **process**.

■ **Spire**
A complete set of **whorls** of a spiral shell.

■ **Sporangiophore**
A spore-producing structure within a **strobilus**.

■ **Squamosal**
A skull bone situated behind the ear in many vertebrates.

■ **Sternite**
The **ventral** plate in an arthropod segment.

■ **Stilt root**
The aerial roots that help support trees such as mangroves.

■ **Stipe**
A branch supporting a colony of individuals.

■ **Striae**
Minute lines, grooves, or channels.

■ **Strobilus (** *pl.* **Strobili)**
A conelike fruiting body present on many land plants consisting of sporangia-bearing structures.

■ **Stromatolite**
A cushionlike, algal growth.

■ **Subchelate**
Having a claw without a fixed finger, and a movable finger operating against a short outgrowth of the hand.

■ **Substrate**
The base on which an animal or plant lives.

■ **Sulcus**
A depression on a shell's surface.

■ **Suspension feeder**
An organism that derives its nourishment from food particles suspended in water.

■ **Suture**
A line on gastropod shells where whorls connect.

■ **Suture (of ammonites)**
The **septum**, often highly convoluted, separating adjacent chambers.

■ **Symbiosis**
The mutually beneficial inter-relationship between two different species.

■ **Symphyseal tooth**
A tooth located in the midline, near the apex of the jaw.

■ **Tabula**
A transverse **septum** that shuts off the lower region of a polyp cavity in some extinct corals.

■ **Tabulate**
Flat, tablelike.

■ **Talonid**
The posterior part of the lower molar tooth in certain mammals.

■ **Tegmen (** *pl.* **Tegmina)**
The thickened forewing of certain insects, such as beetles.

■ **Telson**
In crustaceans, the last segment of the body, containing the anus and/or spines.

■ **Test**
A hard external covering or shell.

■ **Tethys Ocean**
An ancient seaway which stretched from Europe to eastern Asia.

■ **Thallus**
Leaflike structure in primitive plants.

■ **Theca**
In graptolites, the organic-walled tubes housing **zooids**.

■ **Tracheid**
An elongate plant cell with thickened secondary walls.

■ **Tritors**
Specialized dentine in certain fish.

■ **Tube feet**
Tentaclelike structures found in sea urchins.

■ **Tubercle**
A raised mound or bump.

■ **Tumid**
Raised, swollen.

■ **Tympanic bones**
Bones of the ear.

■ **Umbilicus**
The first-formed region of a coiled shell.

■ **Umbo (** *pl.* **Umbones)**
The beaklike first-formed region of a bivalve shell.

■ **Uropod**
A limb on the sixth trunk segment of crustaceans; generally fanlike.

■ **Venter**
In arthropods, the undersurface of the abdomen; in mollusks, the external, convex part of a curved or coiled shell.

■ **Ventral**
Toward the underside.

■ **Water column**
Water depth from surface to bed.

■ **Weberian ossicles**
A chain of three to four bones that connect the swim bladder to the inner ear of some fish.

■ **Whorl**
One complete turn of a shell.

■ **Xiphiplastron**
A paired, bony plate of the lower shell of a turtle.

■ **Zonal marker**
See zone fossil.

■ **Zone fossil**
A fossil species that characterizes strata deposited at a particular time in Earth's history.

■ **Zooid**
An individual of a colonial animal, such as corals, graptolites, and bryozoans.

■ **Zygomatic arch**
In mammals, a bony bar located on the side of the face below the eye.

INDEX

ACKNOWLEDGMENTS

The original edition of this book could not have been completed without access to the magnificent specimens in the collections of the Department of Palaeontology at the Natural History Museum, London, UK and the assistance of both the scientific staff and the members of the photographic unit.

Additional specimens were supplied by Matt Dale, Mr. Wood's Fossils, Edinburgh, Scotland; Mark & Karen Havenstein, Lowcountry Geologic, Charlton, USA; Steve Tracey, London, UK; Vinnie Valle, V&L Crafts, N. Venice, Florida, USA; Hmad and Hamid Segaoui, Segaoui Brothers Erfoud, Morocco and Moha Ouhouiss, Rich, Morocco. Finally, I must thank Charlie Underwood, University of London, UK; Emma Bernard, Natural History Museum, London, UK; Andy Gale, University of Portsmouth, UK and Martin Munt, Dinosaur Isle Museum, Sandown, Isle of Wight, UK for their encouragement, assistance, and advice.

The authors would also like to acknowledge the staff and freelancers (Susie Behar, Peter Cross, Jonathan Metcalf, Mary-Clare Jerram, Gill Della Casa, and Clive Hayball) at Dorling Kindersley, for their enthusiasm, patience, and encouragement that made the original edition so successful. For this edition, the author would like to thank Angeles Gavira at DK London and Hina Jain at DK India for their valuable editorial work and guidance throughout this revision.

Dorling Kindersley would like to thank the following for their work on the original edition of the book: Susie Behar, Jonathan Metcalf, Alison Edmonds, Angeles Gavira, and Andrea Fair for editorial help; Scientific editors Cyril Walker and David Ward; Cyri Peter Cross, Clive Hayball, Elaine Hewson, Alastair Wardle, Kevin Ryan, and Ian Callow for design help; Caroline Webber in production; Contributors Dr. Andy Gale, David Sealey, Dr. Paul Taylor, Dr. Richard Fortey, Dr. Brian Rosen, Sam Morris, Dr. Ed Jarzembowski, Dr. Neville Hollingworth, Steve Tracey, Dr. Chris Duffin, Dr. David Ward, Dr. Angela Milner, Cyril Walker, Andy Current, Miranda Armour-Chelu, Robert Kruszynski, Dr. Jerry Hooker, Mark Crawley; Consultants Dr. Chris Cleal, Dr. Peter Forey, Dr. Andrew Smith, David Sealey; Michael Allaby for compiling the index and glossary; Janos Marffy for airbrush artwork; Will Giles and Sandra Pond for illustrations; Andy Farmer for the illustrations on pp.11, 14, and 15; Adam Moore for computer back-up; Ziggy and Nina at the Right Type; and Caroline Church for Endpaper illustrations.

Dorling Kindersley would like to thank Ankita Gupta and Rishi Bryan for editorial assistance in this edition.

SMITHSONIAN ENTERPRISES
Kealy Gordon, Product Development Manager
Jill Corcoran, Director, Licensed Publishing
Brigid Ferraro, Vice President, Consumer and Education Products
Carol LeBlanc, President

National Museum of Natural History:
Matthew Miller, Collections Volunteer Manager, Department of Paleobiology

PICTURE CREDITS
(Key: a-above; b-below/bottom; c-centre; f-far; l-left; r-right; t-top)

Photography by Colin Keates and David J. Ward except: 13 Dorling Kindersley / Colin Keates: (t). 16 Nature Photographers / Paul Sterry: (cr); M. J. Thomas: (bl). 17 Dorling Kindersley / Colin Keates: (tl); Frank Lane Picture Agency / W. Broadhurst: (tr); E. J. Davis: (cl). 22 The Natural History Museum: (cl); Dorling Kindersley / Colin Keates: (bl). 23 Dorling Kindersley / Colin Keates: (t). 25 Dorling Kindersley / Colin Keates: (tr). 240 Dorling Kindersley / Colin Keates: (tr). 247 Dorling Kindersley / Colin Keates: (br). 255 Dorling Kindersley / Colin Keates: (tr).

Dr. David Ward is a Research Associate based at London's Natural History Museum. His main professional interests are the study of fossil sharks and rays and sieving sediments to extract microvertebrates. He started his career in the early 1970s as a companion-animal veterinarian but switched to paleontology following a major dinosaur collecting trip in west Africa in the late 1980s. Since then, he has led or participated in numerous geological expeditions in Europe, the US, north Africa, and the former Soviet Union. David's passions include education and outreach and bridging the divide between amateur, commercial, and academic paleontologists. He has contributed to more than a hundred scientific papers and received awards for his work from the Geological Society, London, the Palaeontological Association, the Marsh Christian Trust, and the Society of Vertebrate Palaeontology.

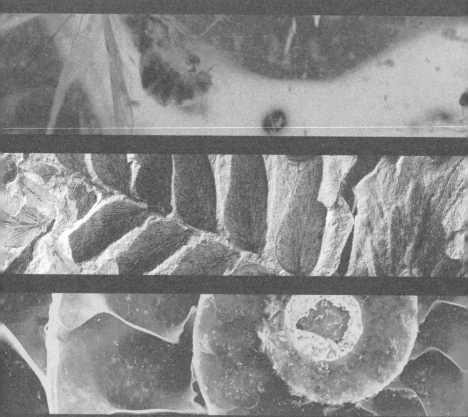